Multiphase Flow Phenomena and Applications

and Applications

Memorial Volume in Honor of Gad Hetsroni

Multiphase Flow Phenomena and Applications

Memorial Volume in Honor of Gad Hetsroni

Editors

Gennady Ziskind

Ben-Gurion University of the Negev, Israel

George Yadigaroglu

ETH-Zurich, Switzerland

 **World Scientific**

NEW JERSEY · LONDON · SINGAPORE · BEIJING · SHANGHAI · HONG KONG · TAIPEI · CHENNAI · TOKYO

Published by

World Scientific Publishing Co. Pte. Ltd.

5 Toh Tuck Link, Singapore 596224

USA office: 27 Warren Street, Suite 401-402, Hackensack, NJ 07601

UK office: 57 Shelton Street, Covent Garden, London WC2H 9HE

British Library Cataloguing-in-Publication Data
A catalogue record for this book is available from the British Library.

MULTIPHASE FLOW PHENOMENA AND APPLICATIONS
Memorial Volume in Honor of Gad Hetsroni

ISBN 978-981-3227-38-5

Printed in Singapore

Preface

This volume is dedicated to a very special person, Professor Gad Hetsroni (1934–2015). His towering figure was familiar to researchers in heat transfer and multiphase flow all over the world. His loss is felt not only by those to whom Gad was close — and there are many people who see him as a friend, or by those to whom he was a mentor — his numerous former students are today leading researchers in academia and industry. In fact, his departure left a void which will never be filled. The only consolation for those who did know him well is the feeling that they were privileged and lucky; others will continue to enjoy the fruits of his active teaching, research and, scientific-community work that spanned more than five decades.

Gad Hetsroni was the founding Editor of the *International Journal of Multiphase Flow* and the person who defined and promoted the discipline around the journal. The unique community formed in this field during his lifetime gathers every three years for a major conference, the *International Conference on Multiphase Flow*, which was held last time in May 2016 in Florence, Italy. This was the first time ever Gad did not attend ICMF. It was the present Editor-in-Chief of the journal, Professor Andrea Prosperetti who immediately proposed to hold Memorial Sessions dedicated to his memory at the conference. We were honored to be asked to organize these sessions; Professor Avram Bar-Cohen joined us and we are grateful for his advice and collaboration. The friends and colleagues from many countries came to Florence to present their personal tributes and scientific papers honoring Gad, while many others attended, marking this touching and memorable event. Our special thanks are going to Professors Alfredo Soldati and Cristian Marchioli, the Conference Chair and Secretary, who also supported us and graciously accommodated three memorial sessions in the very busy conference schedule with about a thousand participants.

Reviewed and edited tributes and scientific papers dedicated to Gad from these Memorial Sessions comprise the core content of this Memorial Volume; certain persons who could not participate in the ICMF made later contributions. We are very happy that the scope of this book can give its readers at least a taste of the enormous variety of Gad's scientific interests, as the contributions deal with gas-solid and liquid-gas multiphase systems, turbulence, hydrodynamics and microscale heat transfer, heat transfer enhancement, and thermal management — the fields to which Gad and his school have made lasting contributions. His former graduate students, postdocs, and research associates are found among the authors, along with his peers from many countries. We hope the tributes from people who have known him well and who were inspired by his personality and ideas will convey the feelings of those who, as said, are lucky to have met him in person. Our special thanks go to Gad's wife, Ruth, and their daughters, Anat and Yael, who generously agreed to contribute a very personal chapter about their beloved father. Yael has also helped us organize the memorial sessions.

Finally, we are very grateful to Dr. Zvi G. Ruder, Senior Executive Publisher of WSPC, who offered to publish this special Memorial Volume and invited us to edit it. We accepted this honor and hope that our modest contribution will serve Gad's blessed memory.

Gennady Ziskind and George Yadigaroglu
February 2017

Contents

Particle-Flow Interaction

Gas-Liquid Two-Phase Flows

Complex Multiphase Systems

Thermal Management on Microscale

Gad Hetsroni: Life Story

Our father and husband, Gad Hetsroni was a man of many traits... his professional life was very impressive, and so was his personal life. We always adored him so much and looked up to him. We miss him so much and want to share some of his life story.

Gad Hetsroni was born on October 10, 1934 in Haifa, Israel, to Kete and Isaac. He had two older brothers. Both parents were very hard working and successful. Kete owned a famous store in Haifa, a model of esthetics and good taste, named "Hetsroni", selling jewelry, silverware, and gifts, while Isaac was a carpenter.

Before the creation of the state of Israel, there were difficult times in Haifa, with much turbulence and uprisings against the Jewish population. In 1938, while Gad was only four, his father was killed in the uprising and his mother was left as the only provider, working long days at the store, leaving Gad with his older brother. His childhood stories included many hours spent together just the two of them, using their lunch money to buy ice-cream.

As a teenager he was literally expelled from school. He was bored and not challenged enough, and was unafraid to show that to the teachers... Knowing where he ended, while all along stressing the importance of education for himself and for us, it is unbelievable to imagine that this was the start of his career, or maybe this explains some of it...

In 1950, at the age of 16, he volunteered, with his mother's permission, to the army where he served in various roles. He participated in all of Israel's wars.

In 1957, Gad graduated with a B.Sc. Cum Laude in Mechanical Engineering from the Technion, Israel Institute of Technology.

In 1959 he met Ruthie, who followed him to the US, and in 1960 they got married, while he was studying for his Ph.D. in Michigan State University. They stayed married and best friends until his last day. We always laughed that in their big house they were always found together in one corner, Ruthie reading her books and Gad working on his computer...

Together they built their family, three daughters, Anat (b. 1961), Orli (1964–1986) and Yael (b. 1967). They have six grandchildren (one granddaughter and five grandsons), named Leor, Yuval, Omer, Stav, Gal, and Sufi. As he said, he always hoped to be surrounded by boys, so all the wooden trucks he bought us girls growing up found good use eventually... All his grandchildren are very close and were very connected to him, as this was of the utmost importance to him. He always wanted to know where everybody is and what they are doing. Family was of utmost importance to him, and to his last day we all knew, daughters, wife, and grandchildren, that he would always be there for us no matter what.

He was always a role model to us in so many ways: he taught us that family is always first, education is of the highest priority, and always told us to do what we want, but learn to be the best we can at it. "If you want, you can be a shoe-maker", he used to tell us, "but study it first, and be the best shoe-maker you can be". He taught us to work hard and excel.

"Did you understand?" he used to ask us, after showing us how to solve our homework in math, physics and science, and after we said "yes" he always threw away the paper and said, "so now you solve it".... He demanded primarily and mostly from himself and then from us.

He was a very generous man, contributing to many causes, especially helping students in their higher education. Just like his family, his students and employees were very important to him, and he was a role model to them as well, being an integral part of their work and research. He always worked with them and not beside them, or as a manager only. His students became family, and the relationships long lasting.

Gad was a man of the world, working hard but also interested in many other fields such as literature, music, and art. He enjoyed good food and wine, along with good company, and loved traveling and exploring the world. Everything was interesting to him, his never-ending curiosity guiding him to learn and do more.

Gad was a very successful scientist, as well as a devoted family man. For us, he was the man who knew everything, curious to find out more, and never tired of educating us. He was always there for us, protecting, loving and showing us the way to excel in any direction we chose. **Just do it** he used to say, to us and himself. A man of few words but many deeds... never tired of working, exploring and enjoying the world.

Dear Dad, we all know how teaching was so important to you... So, we hope you will be glad to know that all of your hard work will live long after you, and will teach and enrich many generations to come. We will keep you as an example of a family man, husband, father, and grandfather, the best anyone could wish for. We were not ready to let you go, and we miss you so much.

Gad Hetsroni and the *International Journal of Multiphase Flow*

Andrea Prosperetti

*Department of Mechanical Engineering,
University of Houston, 4726 Calhoun Rd,
Houston, TX 77204-4006, USA
Department of Applied Sciences and Burgerscentrum,
University of Twente, Enschede, The Netherlands
aprosperetti@uh.edu*

This paper presents an overview of the founding of the *International Journal of Multiphase Flow* and of its development under the editorship of its founder, Professor Gad Hetsroni.

When Gad Hetsroni walked into the office of the 50-year old Robert Maxwell in early 1973, he had in front of him a man who was in the process of restoring his position as head of Pergamon Press after having been expelled from the board in October 1969 on the grounds that "notwithstanding [his] acknowledged abilities and energy, he is not ... a person who can be relied on to exercise proper stewardship of a publicly quoted company." Maxwell was a flamboyant man notorious for his drive, lack of scruples, and ambition, and certainly Gad was apprehensive about the interview. But Maxwell evidently was also a keen judge of character, and he sensed Gad's many abilities and strengths. "When I left," Gad told me, "I had a journal."

This is how the *International Journal of Multiphase Flow* (IJMF in the following) came into existence. The date printed on the cover of Vol. 1, Issue 1 is October 10, 1973, while Issue 2 bears the date of April 16, 1974. The gap of more than 6 months between the first two issues is only apparent: October 10, 1973 was Gad's 39th birthday and Issue 1 was, in fact, published several months later.

The establishment of a new journal in the field of thermo-fluid mechanics was somewhat controversial. The *Journal of Fluid Mechanics*, established by George K. Batchelor in 1956, had already secured its primacy at the top of fluid mechanics journals, and so some felt that a new journal would contribute to the fragmentation of the discipline rather than its unity. Representative of these feelings are the words that Batchelor himself wrote in 1981 (Batchelor, 1981): "A badly chosen field for a new journal will no doubt lead ultimately to its death... Heat and mass transfer, ... multiphase flow, non-Newtonian fluid mechanics, ..., physico-chemical hydrodynamics ... all now have their own journals; is it not scientifically harmful for the minds of readers to be channeled so narrowly?" Batchelor could not foresee the explosion in the scientific literature that took place in the following decades, which has resulted in many of those "narrow" journals (e.g. the *International Journal of Heat and Mass Transfer* and the *Journal of Non-Newtinian Fluid Mechanics*) to thrive. But he was correct in foreseeing the demise of some; for example, *Physicochemical Hydrodynamics*, founded by D.B. Spalding and B.G. Levich, was absorbed by IJMF starting with Vol. 16 in 1990.

Critics did not realize that much of the literature on multiphase flow did not fit the character of the *Journal of Fluid Mechanics* and was itself scattered among various journals devoted to chemical engineering, heat transfer, nuclear safety, the oil industry, and others. Furthermore, the notion that multiphase flow had its own identity was starting to be appreciated in various engineering communities around that time. For example, ASME established the "Polyphase Flow Technical Committee" (itself an offshoot of the Cavitation Technical Committee) in 1970. But it took Gad's intuition and leadership to make the next step and assert multiphase flow as a discipline in itself, beyond and above the boundaries of individual engineering and science specialties. It is not exaggeration to say that Gad "created" the discipline of multiphase flow and coalesced it around "his" journal. As the title page recites — and would recite for many years — "*International Journal of Multiphase Flow* exists for publishing theoretical and experimental investigations of multiphase flow which are of relevance and permanent interest."

The timeliness of Gad's move was responsible for a curious turn of events. Quoting from the Editorial Message at the beginning of Vol. 1, Issue 4, of IJMF: "It is not an uncommon phenomenon in the world of science and engineering for individuals, in different parts of the world, to have similar ideas almost simultaneously; an example of this has been the fact that Professor G. Hetsroni of the Technion in Israel and Professor H. C. Simpson of the University of Strathclyde, in Scotland, in 1972 separately conceived the idea of a journal dealing with multiphase flow phenomena. Professor Hetsroni persuaded Pergamon Press Limited and Professor Simpson persuaded Elsevier Scientific Publishing Company to publish such a journal. The coincidence was carried one stage further in that both editors chose almost identical names for their journals. As soon as the competitive plans became apparent to the two principal editors, discussions were held with a view to combining the two journals. Since this is a rather unusual occurrence in the publishing world, these discussions have been protracted but we are now pleased to announce that the Elsevier *Journal of Multi-Phase Flow* and the Pergamon *International Journal of Multiphase Flow* are now merged." The Editorial Message is followed by a Publishers' Note: "Production and distribution will be in the hands of Pergamon Press Limited to whom all subscription orders should be sent. Three issues of the Pergamon journal had already appeared and the first issue of the Elsevier journal was about to appear when the merger was finalized. This first Elsevier issue has now become the fourth issue of the joint publication, hence its different typography." Indeed, the layout and fonts used for Issue 4 look very different from those used in the preceding and following issues.

The arrangement was successful and continued for a long time. Simpson was listed as co-editor until Issue 4 of Vol. 13 (1987), after which his name appears as a member of the Editorial Advisory Board until issue 5 of Vol. 23 (1997). (He was briefly listed as Associate Editor in the last 2 Issues of Vol. 13 and the first 3 of Vol. 14.) Copyright resided with Pergamon for Issues 4 and 5 of Vol. 1, but was quickly switched to Pergamon and Elsevier, jointly, from Issue 6. In 1991 Pergamon was bought by Elsevier; Pergamon is indicated as the sole copyright owner from Vol. 18 (1992) to Vol. 20 (1994), after which Elsevier became the copyright owner; the imprint was switched to Elsevier only, with no mention of Pergamon, with Vol. 31 in 2005.

When he set up IJMF, Gad was already well-known in spite of his youth and he was able to attract excellent scientists as Associate Editors and as members of the Editorial Advisory Board. The initial Associate Editors were J.A. Bouré (Centre d'Etudes Nucleaires, Grenoble) for basic problems in high-Reynolds-number boiling fluids, H. Brenner (MIT) for laminar multiphase flow, O. Molerus (University of Erlangen-Nürenberg), P.R. Owen (Imperial College), G.F. Hewitt (Atomic Energy Research Establishment, Harwell), and Y. Katto (University of Tokyo) for turbulent multiphase flow, experimental techniques, boiling, and condensation. The Editorial Advisory Board was equally distinguished including the likes of S.G. Bankoff, D. Chisholm, A.E Dukler, H.K. Fauske, P. Griffith, and G.B. Wallis, to name a few.

With his energy, charisma, and organizational skills, Gad had convinced many distinguished colleagues to contribute papers for the critical early issues of IJMF. The first paper (Vol. 1, p. 1) was by D. Barthés-Biesel and A. Acrivos titled "The rheology of suspensions and its relation to phenomenological theories for non-Newtonian fluids." Other contributors included H. Brenner, G.F. Hewitt, G.B. Wallis, and Gad himself. The first few years cannot have been easy. Volume 1 contains 856 pages, but the last issue could not be completed until April 15, 1975. Volume 2 runs for 595 pages and was completed in April 1976, with an issue combining 2 numbers so as to fulfill the editor's promise of 6 numbers per year. Starting with Vol. 3, however, the situation began to stabilize, and Vol. 4 (1978) is the first volume collecting all 6 issues from the same calendar year, as would be the practice from then on. The number of papers published per year oscillates around an upward rising line (Fig. 1), and so does the impact factor (Fig. 2), although the number of pages published per year exhibits stronger oscillations (Fig. 3).

The dashed line in Fig. 3 includes in the page count the pages issued in a supplement titled "Annual Reviews of Multiphase Flow" which Gad launched with Vol. 20 (1994) publishing an additional 416-page issue. The

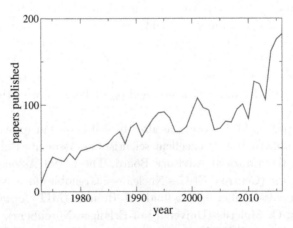

Fig. 1   Papers published per year in IJMF in the period 1973–2016 (papers published in the Supplements in the period 1994–1997 not included).

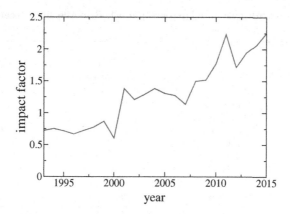

Fig. 2   Impact factor of IJMF since the introduction of this metric in 1993.

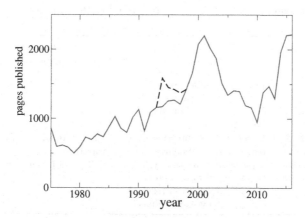

Fig. 3   Pages published per year; the dashed line in the period 1994–1997 includes the pages of the supplements published in this period.

next year, the Supplement ran for only 193 pages, but the initiative did not thrive as it proved very difficult to find enough topics in need of frequent reviews. Gad tried to maintain the Supplement alive by adding lists of multiphase flow papers with abstracts published in other journals, and (with Vol. 23, 1997) even lists of patents, but eventually he had to resign himself and no Supplement was published in 1998 or later.

A look at the most cited papers decade by decade (Tables 1–3, data from Web of Science as of January 31, 2017) gives an interesting glimpse into the evolution of the field and the preoccupations of its practitioners. The first decade is the period in which the early fully resolved simulations,

Table 1   The three most cited papers published in the first decade of IJMF, 1973–1985.

Authors	Title	Volume	Page range	Year	Citations
A. S. Sangani and A. Acrivos	Slow flow past periodic arrays of cylinders with application to heat transfer	8	193–206	1982	401
Y. Sato, Y. Sadatomi and K. Sekoguchi	Momentum and heat-transfer in 2-phase bubble flow. 1. Theory	7	167–177	1981	282
A. S. Sangani and A. Acrivos	Slow flow through a periodic array of spheres with application to heat transfer	8	343–360	1982	271

Table 2   The three most cited papers published in the second decade of IJMF, 1986–1995.

Authors	Title	Volume	Page range	Year	Citations
C. Mundo, M. Sommerfeld and C. Tropea	Droplet-wall collisions — Experimental studies of the deformation and breakup process	21	151–173	1995	436
J. K. Eaton and J. R. Fessler	Preferential concentration of particles by turbulence	20 (Suppl)	169–209	1994	432
M. R. Baer and J. W. Nunziato	A 2-phase mixture theory for the deflagration-to-detonation transition (DDT) in reactive antigranulocytes-materials	12	861–889	1986	425

still limited to fixed particles and Stokes flow, became possible and two such papers (both by Sangani and Acrivos) occupy the first and the third place. The appearance of the paper by Glowinski *et al.* in the second position in the third decade shows the persistent interest in this general topic and the progress made in hardware and software. Air–water flows have endured as a topic of great interest over the entire period. Interest in disperse flow turbulence (to which, we should not forget, Gad made early signal contributions, see e.g. Hetsroni and Sokolov, 1971; Hetsroni, 1989) peaked in the second decade with the advent of new experimental methods and the point-particle model, to reach a period of latency as new computational and experimental methods were being developed.

Table 3 The three most cited papers published in the third decade of IJMF, 1996–2005.

Authors	Title	Volume	Page range	Year	Citations
K. Mishima and T. Hibiki	Some characteristics of air–water two-phase flow in small diameter vertical tubes	22	703–712	1996	477
R. Glowinski, T. W. Pan, T. I. Tesla and D. D. Joseph	A distributed Lagrange multiplier fictitious domain method for particulate flows	20	755–794	1999	461
K. A. Triplett, S. M. Ghiaasiaan, S. I. Abdel-Khalik and D. L. Sadowski	Gas–liquid two-phase flow in microchannels — Part I: Two-phase flow patterns	25	377–394	1999	429

Table 4 Authors with the most articles published in IJMF in the period 1973–2016.

Author	Number of papers
TJ Hanratty	48
G Hetsroni	40
D Barnea	37
DD Jospeh	34
BJ Azzopardi	29
GF Hewitt	28
Y Taitel	27
N Brauner	24
A Mosyak	22
RVA Oliemans	22

Table 4 shows a ranking of the authors with the most articles (excluding editorials, etc.) published in IJMF since inception through the end of 2016. For more than a decade, Gad contributed very little, and actually nothing between 1981 and 1988. He told me that he felt embarrassed by the fact that his associate editors would feel pressured to accept his papers. Probably urged by his younger collaborators at the Technion, he overcame these feelings from the mid-90's until 2006, when he reverted to his previous restraint. In any event, less than one-third of his total production (nearly 150 articles in all) can be found in "his" journal.

With over 100 entries, the Technion is the institution most represented in IJMF, followed by Tel Aviv University and the University of Illinois (about 80 papers each), Delft University of Technology (about 70 papers), and Imperial College (about 60 papers). The country most represented in the journal is the United States, with over 1000 entries, followed by the UK (about 330), Japan, Canada, and France (each with about 220).

During Gad's tenure, that is, until 2007, IJMF published 2152 articles, increasing in number from about 50 to 100 a year. With an acceptance rate of 30%, this means that the number of submissions increased from about 150 to 300 per year, all of which would have been scrutinized by Gad's keen eye. Even with the help of his excellent associate editors, the amount of work must have been substantial, especially in view of the fact that it relied on "snail mail" for much of that period. When, many years ago, I visited Gad in Haifa, he showed me with pride in his beautiful house on a hill slope on Sweden Street the room out of which the journal was produced. This is a memory that I cherish to this day.

And pride indeed he took in "his" journal. He nurtured it with great dedication, far-sightedness, and tender care, occasionally fighting with the publisher over issues such as fonts, the appearance of the printed page, the layout of equations, and others of a similar kind. These aspects mattered to him as much as the quality of the papers, which was always foremost in his mind. He tirelessly worked with his associate editors to make sure that the reputation of the journal remained high and its usefulness to the world's multiphase flow community uncompromised. In these endeavors, he was helped by many prominent scientists who were happy and honored to work with him for the common good as reviewers, associate editors, and members of the Editorial Advisory Board. We, the current members of that community, owe him a great deal as we still enjoy the fruit of his efforts in our professional lives.

Gad was a prominent figure in Israel and on the scene of world science. During his lifetime and after his passing, several tributes were dedicated to him. IJMF published a special issue in 1999 on occasion of his 65th birthday (Vol. 25 Issues 6–7), with a dedication by the editor of the present volume (Yadigaroglu, 1999). An obituary was published in 2015 (Yadigaroglu *et al.*, 2015), and another one can be found on the web page of the American Society of Mechanical Engineers at https://www.asme.org/about-asme/news/asme-news/obituaries. Several other tributes no doubt exist, of which unfortunately I am not aware.

Gad, several IJMF associate editors, and I met at a conference in Istanbul in 2007. He took us all out to dinner one evening and during the meal, much to my surprise, he passed on the baton of IJMF to me. Not accepting was a notion that would never have crossed my mind. But accepting without trepidation was equally impossible. Gad's shoes were very difficult to fill. His good sense, penetrating physical insight, capacity for work, unerring memory for who was doing what, and whom to ask to review which paper — all this could not be equaled.

It has been a privilege and an honor to know him and be counted among his friends. I sorely miss him.

References

G. K. Batchelor, "Preoccupations of a journal editor", *J. Fluid Mech.*, Vol. 106, pp. 1–25, (1981).

G. Hetsroni, "Particles turbulence interaction", *Int. J. Multiphase Flow*, Vol. 15, pp. 735–746, (1989).

G. Hetsroni and M. Sokolov, "Distribution of mass, velocity, and intensity of turbulence in a 2-phase turbulent jet", *J. Appl. Mech.*, Vol. 38, p. 315, (1971).

G. Yadigaroglu, "Gad Hetsroni — A festschrift issue on the occasion of his 65th birthday", *Int. J. Multiphase Flow*, Vol. 25, p. VII, (1999).

G. Yadigaroglu, G. F. Hewitt, Ruthie, Anat and Yael Hetsroni, "Obituary: Gad Hetsroni, Founding Editor of IJMF", *Int. J. Multiphase Flow*, vol. 73, pp. III–IV, (2015).

Gad Hetsroni, 1934–2015

As a close friend of Gad Hetsroni, I had the privilege, a very painful one, to co-edit this volume with Gennady Ziskind. The following remarks are based on the eulogy that I had prepared for the special session honoring him at the International Conference on Multiphase Flows, ICMF, that took place in Florence, the week of May 22–27, 2016. Three special technical sessions were organized at ICMF in memory of Gad. The scientific contributions contained in this volume are based on the papers that colleagues, associates, and friends of Gad presented to honor him at these sessions.

A brilliant career

Gad Hetsroni was born in 1934 in Haifa; he graduated with a B.Sc. cum laude from the Technion in 1957. In 1963, he obtained his Ph.D. from the Michigan State University. He then worked for a short time at the Atomic Power Division of Westinghouse before joining in 1965 the Faculty of the Technion where he remained till his last days.

In 1973, he created the *International Journal of Multiphase Flow*, a major, long-lasting success. Andrea Prosperetti, the current Editor-in-chief of the Journal has written separately on this. Gad Hetsroni also edited the *Handbook of Multiphase Systems*. Clearly the word *Multiphase* is closely associated with Gad.

During his life, Gad kept returning to the US, mainly as a visitor to the Electric Power Research Institute, EPRI, and to the University of California–Santa Barbara and other places. Gad's accomplishments were recognized with multiple honors and national and international distinguished service awards, including the ICMF Senior Award (2010). He has served as Head of the National Council for Research and Development in Israel and as ASME-International Governor as well as Vice President of Region XIII. He was a Fellow of ASME-International. At Technion, he was the holder of the Danciger Chair of Engineering.

Scholar, researcher, scientist, editor, and mostly teacher

I prefer to let others speak here; a few selected quotations from what Gad's numerous friends and associates wrote:

- an international leader in the field of multiphase flows, and a source of inspiration to us all
  Neima Brauner
- among the founders of multiphase flow as a discipline
  Gian Piero Celata
- the invaluable services he has offered to the multiphase flow community for so many years
  Christophe Morel, Nicolas Goreaud, Jean-Marc Delhaye
- his efforts to advance the field of multiphase flows
  Bennett D. Woods, Thomas J. Hanratty
- has promulgated an interest in multiphase flow problems with unusual tenacity, clarity of vision, and single-minded focus
  G.M. Homsy
- should be credited for advancing the area of multiphase flow as a scientific discipline
  M. Tshuva, D. Barnea, Y. Taitel

Editor of the *Int. J. Multiphase Flow* (1973–)

Again, I will let others speak:

- as the Founder and the Editor of the *IJMF*, Gad has marched this field forward at a rapid pace and positioned it at the focus of science and engineering
  S. Polonsky, L. Shemer, D. Barnea
- the *IJMF*, a journal which bears his indelible stamp
  S. Haber, H. Brenner
- his magnificent task at the editorship of the *IJMF* has been decisive for the development and the dissemination of the scientific results of our community
  J. Fabre
- we sincerely appreciate Gad's immense contributions, as editor of the Journal, to the dissemination of the latest results in multiphase flow research throughout the international scientific community
  G. Karimi, M. Kawaji
- his vision and leadership in establishing the IJMF has dramatically advanced the field
  R.T. Lahey
- ... the outstanding *IJMF...* . through his leadership, persistence and devotion ... became a leading journal and a vehicle for new scientific and engineering progress related to multiphase flow
  M. Tshuva, D. Barnea, Y. Taitel

Teacher at Technion and elsewhere

Gad Hetsroni's life has evolved around the Technion (1965–2015). Since 1974, he was holding the Chair of the Danciger Professor of Engineering; he has founded and was the Director of the Multiphase Flow Laboratory and he has served as Dean of Mechanical Engineering and as Head of the Neaman Institute for Advanced Studies.

He has also researched and taught at Stanford University, the University of Minnesota, the University of New South Wales in Australia, and the University of California–Santa Barbara.

The researcher and scientist

Gad is well known for his numerous contributions in several fields, including

- interaction between turbulence and particles

- technique of using infrared radiometry to measure the temperature distribution on the wall of a conduit: the interaction of particles with the wall in dispersed-particle flows; study of the coherent structures
- surfactants to produce a decay of the turbulence and learn about its sources from a different angle
- heat transfer and fluid flow in microchannels that resulted in the book, *Fluid Flow, Heat Transfer and Boiling in Micro-channels*

In his long research career, he has produced about 250 research papers and had 35 doctoral and masters' students. Most of all, he was the teacher of multiphase flows, passing on the basic knowledge on fluid dynamics and multiphase flows and boiling heat transfer to many generations of researchers and scientists/engineers.

The Short Courses on Multiphase Flows at Stanford and in Zurich/Santa Barbara and elsewhere (1979–1984, ...)

Gad Hetsroni was one of the initiators and directors of a very long series of short courses on multiphase flows that started at Stanford and continued (and still continue) at ETH-Zurich and elsewhere in the US. Over two thousand persons participated in these courses.

The family man

This is what his family wrote:

Dear Husband, Dad, and Grandfather

You had a few families: us, the Technion and *The* Journal. You truly loved and contributed to all "families." For us, you were the man who knew everything, curious to find out more, and never tired of educating us. Always there for us, protecting, loving, and showing us the way to excel in any direction we chose. We know how proud you were of every new issue of the Journal that came out, always trying to explain some of it to us... We, your loving family, will keep you as an example of a family man, husband, father, and grandfather, the best anyone could wish for.

We love you

Ruthie, Anat & Yael

A charming citizen of the world

Gad was also a charming citizen of the world: his mischievous/malicious/ playful smile and his great humor enlightened encounters, dinners, meetings... we appreciated his human qualities, caring for friends, love for music, good food, and other good things in life, curiosity for anything new.

The expression "he was a truly good Mensch" seemed to be coined for him.

I will miss his friendship.

George Yadigaroglu

Professor Gad Hetsroni

Professor Gad Hetsroni, my mentor, was born on October 10, 1934 and was 80 when he passed away on March 16, 2015. Just before his death he was happy to celebrate his 80th birthday among friends who honored him by arranging a workshop at the Mechanical Engineering Department at the Technion, addressing heat, mass, and fluid flow problems, topics he dedicated his life to investigate.

The first time I met Gad was in 1967. He was 34 and I was 21. He was a Professor coordinating the Nuclear Engineering Laboratory and I was an undergraduate student. He was known to be a tough, no-nonsense, demanding teacher, and we were all in awe of him. With time, I learned to know him better. Hidden under what appeared to be an impregnable front was a sensitive compassionate person who took care of the people he loved, family, and students alike. His sense of humor was only surpassed by the sense of responsibility and commitment that extended far beyond the period an advisor and advisee spend together. Being one of his first graduate students toward a D.Sc. degree I experienced it first-hand. He was always there when I needed him in my academic life, and, last but, not least, he agreed to serve as Godfather to my first born son.

The awe that I felt as a young person was replaced by affection and deep appreciation.

I decided to begin with a personal note since Professor Hetsroni's seminal, untiring work as one of the most prominent researcher in the field of Multiphase Flow is well known and documented. Gad possessed a combination of tangible and intangible qualities rarely found in one researcher: The ability to pinpoint what state-of-the-art subjects are worth investigating; what are the experiments needed to unravel the complexities of the problems under investigation; how to design ingenious experiments; how to convince people to support the research and raise sufficient grant money; how to motivate young students to participate in the research; how to summarize the results in a convincing manner, publishable in prestigious journals; how to present the research and its finding in an interesting and intriguing way that made him a most desirable speaker.

Gad founded the prestigious *International Journal of Multiphase Flow* when he was in his thirties and served as its Chief Editor for 35 years. He wrote or co-wrote nine books and more than 150 articles, mainly in the fields of heat transfer and two phase flow. He had been invited to present keynote speeches and seminars all over the globe and served on the editorial boards and organizing committees of numerous conferences.

What was humiliating to us all, quite often younger people, was that at an age he could retire peacefully and enjoy his laurels, he would still arrive first and leave his office last. After the official age of retirement, he still managed the biggest research group in our Department that comprised six mature scientists and 10 graduate students. He tried persistently and tire-lessly to raise grant money, not an easy task in the Israeli environment, and continued travelling to present his findings. His energy and productiveness had not diminished over the years. To the contrary, defying the Laws of Nature, his energy appeared to be growing with time, and the number of papers he published after retirement exceeded that before retirement.

He was and still is a role-model to all of us both in character and in his zeal to do research. The candle he lit in the hearts and minds of his many advisees is still burning, illuminating new fields of research.

May He Rest in Peace

Professor Shimon Haber
Department of Mechanical Engineering
Technion-Israel Institute of Technology

Gad Hetsroni, 1934–2015

Sadly, I was unable to attend the 2016 ICMF in Florence, but I was pleased to hear that some sessions at this Conference had been devoted to the memory of Gad Hetsroni and that the papers from these Sessions had been collected in this Memorial Volume. As founder and first Editor-in-Chief of the International Journal of Multiphase Flow, Professor Hetsroni has had a very strong influence on the establishment of the multiphase flow field and he must have been very proud of the way the field has developed over the past four decades.

I first met Gad Hetsroni in the early 1970s when I visited Haifa and have been in constant touch with him ever since. In addition to interacting for many years on the journal, we were involved in teaching on short courses of which he was organizer. The first such course was held at Stanford University in 1979, and subsequent offerings of the course were made at UC Santa Barbara and (under the leadership of Professor George Yadigaroglu) at ETH Zurich. Over 2000 students have attended these courses and (until he passed away) Professor Hetsroni always gave the opening lecture! One excellent feature of these courses has been the associated social life, and I remember well having many enjoyable dinners with Gad and his wife Ruth, who was very supportive to him in these and many other endeavors.

Gad Hetsroni has been such a part of my life that it seems strange to now be without him! However, I guess that is part of life and one has to accept it. However, his legacy in our field will persist into the future and outlive all of us!

Geoff Hewitt
08.07.16

My Encounters with Gad Hetsroni*

Gennady Ziskind

*Heat Transfer Laboratory, Department of Mechanical Engineering
Ben-Gurion University of the Negev, Beer-Sheva, Israel*

Although for the first time I probably met Gad Hetsroni in late 1990 or early 1991, it happened that I never attended his class during my graduate studies, and my real acquaintance with him occurred in the summer of 1995, when he hired me as the research engineer for the Energy Research Center that he headed at that time. He was already really famous then and, in fact, approaching the age when many people are retiring ...

Fast forward to November 25, 2014: many of us gathered that day at the Technion for the celebration of Gad's 80th birthday, organized by Shimon Haber. Among the speakers, there was Technion's President, Professor Peretz Lavie. His speech combined high praise of Gad's work with some — apparently unavoidable in our times — metrics, like the number of papers Gad had published or citations of his works. The President expressed some amazement of Gad's continuing research work and steady output. Indeed, during his "retirement years" Gad Hetsroni produced a huge amount of high-quality research papers, applied for and received multiple grants, and supervised graduate students. So, all that was in no sense an exaggeration, which quite understandably happens sometimes on occasions of this sort. However, I was sitting there with a strange feeling: somehow Gad had always known that would happen. Back in the mid-1990s, we worked on some proposals, and one of the requirements was to append a complete CV of the Principal Investigator. So, I came to his office and was given a thick pack of paper probably requiring a very special stapler. While I was looking

---

*Based on a presentation "Gad Hetsroni: Some Reminiscences of Him in My Life and Work, From 1995 to This Day and Beyond", delivered at ICMF in May 2016.

at it in bewilderment, Gad said without any emotion: "I am always telling my daughters this only depends on your age". This, it turns out, was not only an assertion but a prediction as well.

People who knew Gad are well aware of his dry wit, which could be rather caustic at times. I remember well how he, in a phone conversation, was trying to persuade a technical support engineer that his infrared camera was giving wrong readings. After a couple of rounds the guy apparently did not give in, so Gad told him rather matter-of-factly that the fact that you need to have a surface at above 100°C to boil water at atmospheric pressure had been established by Claude Debussy. In fact, however, Gad was extremely attentive and caring with people who worked in his group, especially if they had some disadvantage or were put in some unusual and pressing circumstances. I remember quite well how he brought his wife, Ruth, to give Hebrew lessons to people who had recently arrived in Israel and for whom those lessons were a sort of lifeline.

These are only flashbacks about his really towering figure, from those wonderful 20 years ...

1995

- Still in my Ph.D. with Chaim Gutfinger when Gad hires me as a research engineer, "half-engineer" in his definition (in response to a clever decision by the administration to split the position between two persons)
- I start to work with Albert Mosyak and Leonid Yarin on almost everything related to his lab

1996

- We start to work on the first proposal on microchannel heat transfer

- Among other things, it started the brilliant career of Yoav Peles
- This is what it was in Gad's handwriting

 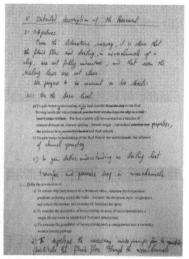

1997

- Gad is trying to find a postdoc for me

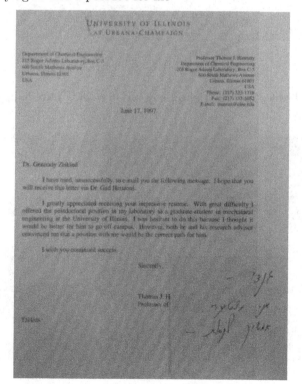

- He writes about a negative response:
  Gennady
  I am sorry
  I shall continue to try

1998

- I am going to BGU to present a seminar as a candidate...
Gad gives me his go-ahead to talk about the work done in his lab.

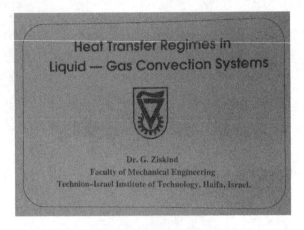

1999

- With Gad's blessing, I move to Ben-Gurion where I have happily dwelled since then...

2000–2002

- We work on a paper dealing with boiling in microchannels
- The way wasn't easy

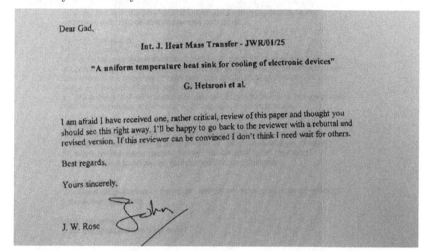

Dear Gad,

Int. J. Heat Mass Transfer - JWR/01/25

"A uniform temperature heat sink for cooling of electronic devices"

G. Hetsroni et al.

I am afraid I have received one, rather critical, review of this paper and thought you should see this right away. I'll be happy to go back to the reviewer with a rebuttal and revised version. If this reviewer can be convinced I don't think I need wait for others.

Best regards,

Yours sincerely,

J. W. Rose

PERGAMON   International Journal of Heat and Mass Transfer 45 (2002) 3275–3286

International Journal of
HEAT and MASS TRANSFER

www.elsevier.com/locate/ijhmt

A uniform temperature heat sink for cooling of electronic devices

G. Hetsroni [a,*], A. Mosyak [a], Z. Segal [a], G. Ziskind [b]

[a] *Department of Mechanical Engineering, Technion-Israel Institute of Technology, Haifa 32000, Israel*
[b] *Department of Mechanical Engineering, Ben-Gurion University of the Negev, Beer-Sheva 84105, Israel*

Received 13 March 2001; received in revised form 28 January 2002

- But few years later it turns out that ... it had not been that bad:

Dear G. Hetsronoi,

I am pleased to inform you, as corresponding author, that your article A uniform temperature heat sink for cooling of electronic devices, G. Hetsroni, A. Mosyak, Z. Segal, G. Ziskind, Volume 45 (2002), Issue 16, Pages 3275-3286 is one of the most cited articles from for the years 2002 to 2005*.

The popularity of your work suggests that it is of particular interest and value to researchers in the field, and we are delighted that you chose to publish it in our journal.

2006–2009

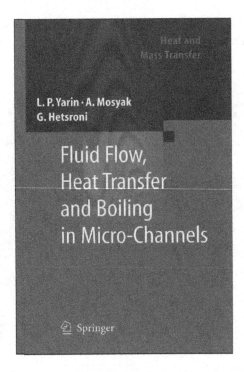

- The book by Yarin, Mosyak, and Hetsroni is in works
- I am happy they found it possible to ask for my opinion

Letter about Chapter 4
Gennady Ziskind [gziskind@bgu.ac.il]

Sent:	Sunday, November 11, 2007 2:36 PM
To:	Albert Mosyak [mealbmo@techunix.technion.ac.il]
Bcc:	גנדי זיסקינד
Attachments:	Pismo.tif (346 KB)

- I found myself the only person mentioned twice:

We would like to express our gratitude to our colleagues Professor J. Zakin, Dr. E. Pogrebnyak, Dr. R. Rozenblit, Dr. Z. Segal, Dr. I. Tiselj, Dr. G. Ziskind, as well as D. D. Klein (M.Sc.), Y. Mishan (M.Sc.) and R. Zimmerman (M.Sc.) for their participation in the investigations of a number of problems considered in this book.

We are especially grateful to Dr. M. Fichman and Dr. G. Ziskind for many valuable discussions and comments made after reading the manuscript.

- As one of Gad's favorite sayings goes:
Two are better than one

2012

- Gad hosts four of my graduate students and myself in his lab. He enquires about them:

From: Gennady Ziskind [gziskind@bgu.ac.il]
Sent: Wednesday, August 08, 2012 5:17 PM
To: Prof. Hetsroni
Subject: Re: RE: thank you
Havatzelet is a Ph.D. student
Yoram is almost done with his M.Sc. thesis and will start
his Ph.D. in October
Liel is an M.Sc. student
Tomer completed his third year and starts his direct-
track M.Sc. thesis now

- Three of them have submitted their Ph.D. theses since then, the fourth is in works

2013

- Gad invites me to submit a joint proposal

----- Original Message -----
From: Gad Hetsroni <hetsroni@tx.technion.ac.il>
Date: Monday, April 29, 2013 8:25
Subject: FW: Joint IIT- BGU Call for Proposal
To: Gennady Ziskind <gziskind@bgu.ac.il>

Pre-proposal to Technion & BGU

Thermal Management of Concentrating Photovoltaic Cells

G. Hetsroni*, G. Ziskind** and G. Yossifon*

* Technion – Israel Institute of Technology, Dept. of Mechanical Engineering, Haifa.

Phone: 04-829-2058 Fáx: 04-823-8101 E-mail: hetsroni@tx.technion.ac.il

**Associate Professor and Head, Heat Transfer Laboratory

Ben-Gurion University of the Negev, Dept. of Mechanical Engineering, Beer-Sheva.

Phone: 08-647-7089 Fax: 08-647-2813 E-mail: gziskind@bgu.ac.il

2014

- Celebration of Gad's 80th birthday at the Technion

יום עיון לכבוד יום הולדתו ה 80 של פרופסור גד חצרוני 25.11.14			
09:30-10:00	התכנסות		
	(חדר 217 בניין דן קאהן - כיבד קל)		
10:00-10:20	ברכות		
	Introduction	Shimon	Haber
	Dean of Mechanical Engineering address	Yoram	Halevi
	President of Technion address	Peretz	Lavie
10:20-12:30	אנרגיה		
10:20	ynote lecture-The future of nuclear energy אטומית לאנרגיה הוועדה ראש יעבר	Gideon	Frank
11:00	Flow and mass transfer in micro channels	Rafi	Semiat
11:30	Close-contact melting in geometries suitable for latent heat storage	Genady	Ziskind
12:00	A Coanda-based reciprocating wind energy generator	David	Greenblat
12:45-14:00	ארוחת צהרים		
	(מרצים+משפחת חצרוני ואורחיה VIP מועדון הסגל בחדר)		
14:15-15:45	חלקיקים בזרימה		
14:15	Microswimmers	Shimon	Haber
14:45	Measurement of forces on Lagrangian particles in turbulent flows	Alex	Liberzon
15:15	Virtual vials for enhanced biomolecular analysis	Moran	Bercovici
15:45-16:00	נעילה		
15:45	daughters+Gad Hetsroni's closing address	Anat	Hetsroni

2016 and beyond

- Gad Hetsroni Memorial Sessions take place at ICMF in Florence

- This Memorial Volume is being prepared and published
- My new role as an Israeli delegate to AIHTC (Assembly of International Heat Transfer Conferences) starts — entering Gad's big shoes ... or, rather,

standing on the shoulders of giants ...

Tribute to Prof. Gad Hetsroni

Masahiro Kawaji

Professor of Mechanical Engineering, City College of New York,
New York, NY 10031, USA
and Associate Director of the CUNY Energy Institute

We are grateful for the support and encouragement for multiphase flow research provided by Prof. Gad Hetsroni both personally and as the founding editor of the *International Journal of Multiphase Flow*. His contributions to the growth of multiphase flow research all over the world will long be remembered and appreciated.

Extension of POD Analysis Toward the Three-Dimensional Coherent Structure

R. Gurka[*,§], A. Liberzon[†] and G. Hetsroni[‡]

*School of Coastal and Marine Systems Science,
Coastal Carolina University, Conway, SC, USA
†School of Mechanical Engineering,
Tel Aviv University, Tel Aviv, Israel
‡Faculty of Mechanical Engineering,
Technion, Haifa, Israel
rgurka@coastal.edu

We summarize in this contribution our quest toward characterization of a 3D coherent structure. The study started with the microbubble flow visualization experiments of Gad Hetsroni and his co-authors back in the 1990s and expanded during 1999–2003 study using particle image velocimetry (PIV) and proper orthogonal decomposition (POD). The topology of large scale structures in the turbulent boundary layer in a flume is investigated experimentally. The 2D velocity fields are obtained by a stereoscopic PIV system that enables to measure out-of-plane velocity components. The large scale structure is reconstructed from the POD modes of vorticity in three orthogonal planes. Extending on the previous studies, the dominant coherent pattern is visualized in 3D, combining the three orthogonal projections. The pattern appears as a streamwise oriented vortex structure of length of ≈ 1000 wall units and with the inclination angle of $\approx 9°$. The proposed POD-based characterization method leads to a conceptual illustration of a 3D structure. The shape resembles the "funnel" structure proposed by Prof. Hetsroni and colleagues in 1990s based on the microbubble flow visualization images.

---

§This paper is dedicated to the memory of Gad Hetsroni. His inspiration, vision and mentorship guided us during our PhD studies and throughout our academic career. Gad was an inspirational leader and educator.

1. Introduction

Turbulent boundary layers are responsible for momentum or heat transfer, turbulent kinetic energy production, and dissipation in wall bounded flows. Large coherent structures are relevant for the large portion of the turbulent transport [Baltzer *et al.*, 2013; Kaftori *et al.*, 1994; Kline *et al.*, 1967; Li and Bou-Zeid, 2011; Marusic, 2001; Panton, 1997]. Despite extensive research, the precise role of coherent structures in the turbulent boundary layer mechanisms and their relation to vorticity, Reynolds stress, energy production, and dissipation are unresolved [Tsinober, 2000]. Furthermore, there is no consensus regarding their size and topology between the hairpin packet model [Adrian, 2007; Smith and Walker, 1995; Zhou *et al.*, 1999], horseshoe vortex [Theodorsen, 1952], "funnel" [Kaftori *et al.*, 1994], near-wall longitudinal vortices [Schoppa and Hussain, 2000a], vortex clusters [Lozano-Durán and Jiménez, 2014], or very large-scale structures [Lee *et al.*, 2014; Smits *et al.*, 2011]. One of the main challenges is their 3D and time-dependent nature, see for reviews McNaughton and Brunet [2002]; Panton [1997]; Robinson [1991]. There is, however, some similarity reported of a mean angle in the streamwise direction of approximately 8°/12° [Adrian, 2007; Zhou *et al.*, 1999], which is in a good agreement with the experimental observations of Kaftori *et al.* [1994].

In this study, we extend the characterization method reported in Gurka *et al.* [2006]. The method, based on the proper orthogonal decomposition (POD) [Berkooz *et al.*, 1993; Holmes *et al.*, 1996; Liberzon *et al.*, 2001; Lumley, 1970] provided a statistical view on the dominant coherent structures. Gurka *et al.* [2006] demonstrated that using a linear combination of POD modes of vorticity fields, it is possible to characterize geometrical features of coherent motions. Thus, the authors obtained the angle and the length in the streamwise-normal plane of a large coherent structure within the boundary layer. The streamwise-normal plane was reconstructed in Gurka *et al.* [2006]. Recently, we combined the 2D measurements with the 3D data obtained using our multiplane stereoscopic PIV [Liberzon *et al.*, 2004, abbreviated XPIV] and have shown the direct comparison of the reconstruction using either a set of 2D POD modes or the directly calculated 3D POD modes in a turbulent boundary layer [Liberzon *et al.*, 2011]. The analysis has shown that the key topological features of the large scale flow pattern are reconstructed equally well by both methods. Based on these studies, we can now extend our analysis to the characterization of a coherent motion in 3D if we combine POD modes measured separately in the orthogonal planes. Using this reconstruction, we can propose a plausible illustrative description of a 3D flow pattern that resembles the aforementioned model of "funnels".

This contribution is organized as follows. In Sec. 2, we briefly review the experimental setup and introduce the POD snapshot method applied to the measured vorticity fields. In Sec. 3, we present the central result of the 3D reconstruction, followed by an illustrative model presented in Sec. 4.

2. Methods

2.1. *Proper Orthogonal Decomposition (POD)*

POD extracts coherent structures based on orthogonal eigenfunctions of Karhunen–Loéve decomposition [Berkooz *et al.*, 1993; Holmes *et al.*, 1996; Lumley, 1970], following the solution of a Fredholm integral equation:

$$\int R_{ij}(x,x')\phi_j(x')dx' = \lambda\phi_i(x), \tag{1}$$

where $R_{ij}(x,x')$ is the two-point correlation matrix of N realizations of the random field (in our case it is vorticity ω)

$$R_{ij}(x,x') = \frac{1}{N}\sum_{i=1}^{N}\omega_i(x)\omega_j(x'). \tag{2}$$

A low-order model of the random field $\hat{\omega}(x)$ is obtained by reconstruction using only first M strongest eigenmodes

$$\widehat{\omega}_i(x) = \sum_{n=1}^{M}\lambda^{(n)}\phi^{(n)}(x) \qquad i = 1,2,3. \tag{3}$$

A computationally efficient "method of snapshots" [Sirovich, 1987] is used to compute the decomposition and reconstruction of the vorticity fields. The vorticity is more suitable for the POD analysis as we adopt the definition of the vortical structure according to a *large-scale turbulent fluid mass with spatially phase-correlated vorticity*, Hussain [1986].

Similarly to the classical work of Lumley [1970] and later studies like Gordeyev and Thomas [2002]; Gurka *et al.* [2006]; Liberzon *et al.* [2011], we use in Eq. (3) a linear combination of the dominant POD modes as a description of the term "large scale structure". The spatial localized coherent structures appear as streamwise elongated regions of the condensed vorticity that do not change their topology by combining $M = 3, 5, 10$, or 150 POD vorticity modes [Gurka *et al.*, 2006; Liberzon *et al.*, 2011]. They shown that the representation of the large scale structure using the first three vorticity POD modes is sufficient to define its topology, lengths, and an inclination angle.

2.2. Experimental setup

The turbulent boundary layer of an open flume made of glass ($4.9 \times 0.32 \times 0.1$ m) was measured in three orthogonal planes: streamwise–wall normal $x-y$, streamwise-spanwise $x-z$, and wall normal-spanwise $y-z$ (Fig. 1) and for eight different conditions (at $x = 2.5$ m from the entrance) at Reynolds numbers ranging between 21,000 and 57,000, where Reynolds is based on the water height in the flume and the mean velocity. More details on the flume and the experimental setup may be found in Liberzon *et al.* [2004].

A standard stereoscopic PIV system (TSI Inc.) was used to measure the flow field providing a 2D three velocity components in a cross-section within the boundary layer. Hollow glass sphere particles (mean diameter 11 μm) were introduced to the flow at the downstream tank to allow mixing. We used 64×64 pixels (80 μm/pixel magnification ratio) interrogation windows

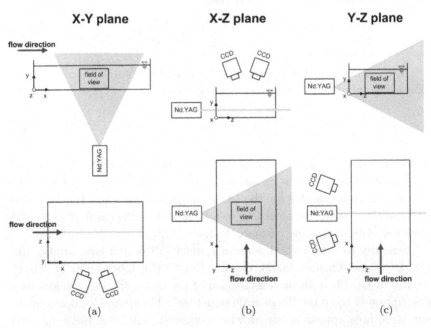

Fig. 1 Experimental configuration setup: streamwise-wall normal plane, $x-y$ (a), streamwise-spanwise, $x-z$ plane (b), wall normal-spanwise, $y-z$ plane (right).

and 50% overlap, producing about 1000 vectors per map. Data was filtered using standard median and global outlier filters, with 5% of the erroneous vectors removed during the post-processing analysis. Image processing and PIV analysis were performed using commercial Insight 3G software (TSI Inc.) and verified using an open source OpenPIV software [Taylor *et al.*, 2010]. The software also has the POD toolbox used for the present analysis.

3. Results

The large-scale structure was reconstructed from the POD eigenmodes, based on the 2D projections of the instantaneous vorticity field realizations, in three orthogonal planes. The utilization of POD on the measured data in the flume was for eight flow conditions based on the Reynolds number [see Gurka *et al.*, 2006, for further details]. We used the linear combination of the first three POD modes according to Eq. (3), to identify the large scale structures and provided a qualitative description on the mechanisms associated with them. Figure 2 depicts a "visual reconstruction" of the combined POD modes for the three orthogonal planes as color shaded maps. The $x-y$ plane was analyzed in detail and presented in Gurka *et al.* [2006]

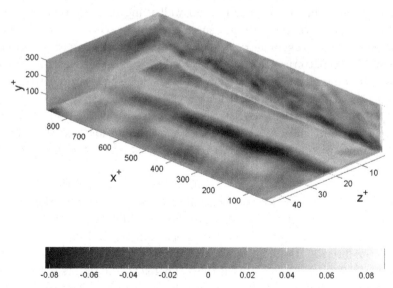

Fig. 2   3D reconstruction of coherent patterns in three orthogonal planes, $x-y$, $x-z$ and $y-z$, using color shaded fields of reconstructed vorticity $\widehat{\omega}$. The flow is in the positive direction of x (from right to left), Re = 27,000.

for a range of Reynolds numbers, decomposing both velocity and vorticity based on the number of modes required to reconstruct the original datasets which later on were used to represent the flow structures in the flume.

The two coherent patterns in the $x-y$ plane show a projection of an elongated vortical structure, inclined and expanded in the streamwise direction. It is noteworthy that in this plane we estimate the spanwise vorticity component of a 3D vortical structure, and it should be seen as a projection of the 3D vorticity vector onto the $x-y$ plane. It implies that these coherent patterns are the projection of a quasi-streamwise structure, elongated in the streamwise direction, and inclined upward from the wall. The patterns appear in the near wall region ($20 < y^+ < 100$), their non-dimensional streamwise extension is estimated to be in the range of $600 < x^+ < 1000$ and the average inclination angle of $\approx 9°$. The patterns match the description of bursting events, consistent with the classical models of Schoppa and Hussain [2000b] and observed in the $x-y$ plane in the measurements of Bogard and Tiederman [1986].

The patterns visible in the combination of the POD modes of wall normal vorticity in a plane parallel to the bottom of the flume ($x-z$ plane) resemble the low-speed streaks [Kline et al., 1967] as shown in Fig. 2. We have observed such structures in Gurka et al. [2004] measuring the thermal signatures directly on the bottom wall using infrared thermography in conjunction with the PIV measurements in the $x-z$ plane. The pattern corresponds to symmetric, elongated positive, and negative vorticity regions. These patterns show similarity to the low-speed streaks due to vortical structures with typical ejections from the low-speed region at the wall upward toward the mean flow [Klewicki, 1997; Smith, 1984]. It is noteworthy that the average distance between these patterns at the wall is approximately 100 wall units ($z^+ \approx 100$), and they extend for several hundreds of wall units in the streamwise direction ($600 \le x^+ \le 1000$), repeatedly reported since Kline et al. [1967].

The $y-z$ plane was the most difficult measurement setup, since the investigated plane was located normal to the mean flow direction, which imposed a strong out-of-plane motion [Fei and Merzkirch, 2004; Willert, 1997]. The stereoscopic PIV measurements inherently have larger error, and after filtering, result in lower ratio of valid stereo PIV data [Fei and Merzkirch, 2004; Lawson and Wu, 1997]. Therefore, the results in this plane have lower resolution as compared with the other planes. Nevertheless, the projection of coherent patterns in the reconstruction on this plane resembles streamwise vortical structures [e.g. Jeong et al., 1997]. These structures also fit the model proposed by Brooke and Hanratty [1993], in which an opposite signed offspring vortex forms immediately underneath a parent vortex, whose downstream end is lifted from the wall. The upper vortex was coupled with a counter-rotating vortex near the bottom. The lower vortex was mainly governed by the interaction with the bottom, meaning the low speed streaks.

Note that the $x-z$ plane represents a cross section at $y^+ = 130$ and not at the bottom wall. We assume that the mean velocity in the turbulent boundary layer is in steady state as well as that the POD modes represent the some sort of an energetic (in terms of vorticity) level corresponding to the observed structures. It is therefore not surprising that Fig. 2 shows a remarkable similarity and visual "connection" between the three orthogonal planes. Recalling that each plane shows a different vorticity component of the 3D vorticity vector, we can imagine the 3D structure that would create such projections as a swirling quasi-streamwise vortical structure, inclined upward and expanding along the streamwise direction.

4. Discussion

In this section, we present our experimental results obtained and characterized using PIV and POD in three orthogonal planes in light of the coherent structure models. Many researchers attempted to characterize the coherent motions, governing the flow, by describing them as vortex structures, reviewed by Robinson [1991] and other works cited here.

The most common illustration of a vortex is a swirled fluid motion, in some sense similar to a moving tornado or whirlpool laying on a side and rotating around a streamwise oriented axis. Turbulence is illustrated as being made up of a number of vortices that move around, and interact. Flow visualization techniques emphasized different characteristics associated with coherent structures, e.g. low-speed streaks and bursting phenomena in a turbulent boundary layer or von Karman vortices in wakes of bluff bodies. Similar to von Karman vortices, depicted as coherent structures of organized rollers at the wake region based on the vorticity vector, our results are based on vorticity field analysis, obtained either by numerical simulations [Liberzon *et al.*, 2005] or experiments [Gurka *et al.*, 2004, 2006; Liberzon *et al.*, 2001, 2011].

The proposed conceptual model is based on the model proposed by Kaftori *et al.* [1994]. Based on the flow visualization methods, the authors suggested that the structures have a shape of an expanding spiral vortex shaped like a "funnel". The structure is laid sideways in the streamwise

(a)   (b)

Fig. 3   Conceptual illustration of coherent structure in a turbulent boundary layer: (a) a vortical loop shape structure, (b) group of vortical loop shape structures.

direction with its "head" starting in the near wall region and the funnel "base" expanded in three dimensions downstream. Figure 3(a) is an illustrative description of such structure, shown by swirling vortex, which marks an isosurface of vorticity magnitude. Such a vortical pattern is sheared in the turbulent boundary layer flow and has strong spanwise and wall normal vorticity components in addition to the main streamwise vorticity. Such a coherent motion can be seen as a vortical loop shape which is both tangled and expanded in the streamwise direction. We suggest two major mechanisms in this conceptual model. One describes a Lagrangian path of a fluid element that experiences stresses and deformations during its movement. The second is an imaginary envelope, which is somewhat similar to the concept suggested by McNaughton and Brunet [2002], that controls the transport mechanisms such as: momentum, energy, and heat. Another aspect of the proposed model is the three-dimensional nature of the pattern that presents strong vorticity projections in the three orthogonal planes. The turbulent transport is further enhanced due to the presence of similar neighborhood structures, which have strong interactions with each other. An illustrated group of structures is sketched in Fig. 3(b), which gives a description of a group of structures.

It is noteworthy that Fig. 3(b) does not necessarily represent an instantaneous event of coexistence of multiple structures. It can be expected that the strong interactions among the structures lead to their destruction or "coalescence". Yet, we may assume that these are statistically dominant types of structures that can appear as single or multiple patterns within the flow.

The suggested model of coherent motions resembles many models available in the literature, such as Blackwelder and Eckelmann [1979]; Nezu and Nakagawa [1993]; Schoppa and Hussain [2000a], among others.

Similar to all other models, the geometrical relations could be described via our model, including an inclination angle ($\sim 10°$), non-dimensional length (~ 1000), and width (~ 100) in wall units. The conceptual model is different compared with others [Zhong *et al.*, 2016] as it considers the strong interaction between the flow patterns within the boundary layer.

It is noteworthy that the model describes a large structure in the turbulent boundary layer. The resolution of the low-order representation is not suitable for characterization of the small scale structures such as hairpins or hairpin packets [Adrian, 2007; Zhou *et al.*, 1999], nor does it fit to fully-identify the very large coherent structures that were presented more recently [Smits *et al.*, 2011] where high-Reynolds numbers are accounted.

5. Summary

This study of coherent structures in a zero-pressure-gradient turbulent boundary layer is based on measurements of the three-component, 2D velocity fields by using stereoscopic PIV technique. The measurements were performed in a flume in three orthogonal planes. The statistical description of the 2D footprints of the coherent motions was obtained by means of POD of the vorticity components. The imaginary combination of the orthogonal plane footprints in the 3D view proposes the existence of the large scale, quasi-streamwise structure, elongated in the streamwise direction. The determined topology based on our identification procedure can be a useful tool in controlling the turbulence mechanisms, which affect engineering and science applications.

Acknowledgments

This contribution is devoted solely to the memory of Prof. Gad Hetsroni (1934–2015) who was our Ph.D. adviser during 1999–2003.

References

R. J. Adrian, "Hairpin vortex organization in wall turbulence", *Physics of Fluids*, Vol. 19, No. 4, p. 041301, (2007).

J. R. Baltzer, R. J. Adrian and X. Wu, "Structural organization of large and very large scales in turbulent pipe flow simulation", *J. Fluid Mech.*, Vol. 720, pp. 236–279, (2013).

G. Berkooz, P. Holmes and J. L. Lumley, "The proper orthogonal decomposition in the analysis of turbulent flows", *Ann. Rev. Fluid Mech.*, Vol. 25, pp. 539–576, (1993).

R. Blackwelder and H. Eckelmann, "Streamwise vortices associated with the bursting phenomena", *J. Fluid Mech.*, Vol. 94, No. 3, pp. 577–594, (1979).

D. Bogard and W. Tiederman, "Burst detection with single-point velocity measurement", *J. Fluid Mech.*, Vol. 162, p. 382, (1986).

J. Brooke and T. Hanratty, "Origin of turbulence-producing eddies in a channel flow", *Phys. Fluids*, Vol. A No. 5, (1993).

R. Fei and W. Merzkirch, "Investigations of the measurement accuracy of stereo particle image velocimetry", *Experiments in Fluids*, Vol. 37, No. 4, pp. 559–565, (2004).

S. Gordeyev and F. Thomas, "Coherent structure in the turbulent planar jet. Part 2. Structural topology via POD eigenmode projection", *J. Fluid Mech.*, Vol. 160, pp. 349–380, (2002).

R. Gurka, A. Liberzon and G. Hetsroni, "Detecting coherent patterns in a flume by using PIV and IR imaging techniques", *Experiments in Fluids*, Vol. 37 No. 2, pp. 230–236, (2004).

R. Gurka, A. Liberzon and G. Hetsroni, "POD of vorticity fields: a method for spatial characterization of coherent structures", *Int. J. Heat Fluid Flow*, Vol. 27, No. 3, pp. 416–423, (2006).

P. Holmes, J. Lumley and G. Berkooz, *Turbulence, Coherent Structures, Dynamical Systems and Symmetry*, Monographs on Mechanics, Cambridge University Press, Cambridge 1996.

A. Hussain, "Coherent structures and turbulence", *J. Fluid Mech.*, Vol. 173, pp. 303–356, (1986).

J. Jeong, F. Hussain, W. Schoppa and J. Kim, "Coherent structures near the wall in a turbulent channel flow", *J. Fluid Mech.*, Vol. 332, pp. 185–214, (1997).

D. Kaftori, G. Hetsroni and S. Banerjee, "Funnel-shaped vortical structure in wall turbulence", *Phys. Fluids*, Vol. 6, pp. 3035–3050, (1994).

J. Klewicki, "Self-sustaining traits of near-wall motions underlying boundary layer stress transport", *Self-Sustaining Mechanisms of Wall Turbulence*, R. L. Panton (Ed.), Advances in Fluid Mechanics, Chapter 7, Computational Mechanics Publications, UK, (pp. 135–166), (1997).

S. Kline, W. Reynolds, F. Schraub and P. Runstadler, "The structure of turbulent boundary layers", *J. Fluid Mech.*, Vol. 30, pp. 741–773, (1967).

N. Lawson and J. Wu, "Three-dimensional particle image velocimetry: error analysis of stereoscopic techniques", *Meas. Sci. Tech.*, Vol. 8, pp. 894–900, (1997).

J. Lee, J. H. Lee, J. I. Choi and H. J. Sung, "Spatial organization of large-and very-large-scale motions in a turbulent channel flow", *J. Fluid Mech.*, Vol. 749, pp. 818–840, (2014).

D. Li and E. Bou-Zeid, "Coherent structures and the dissimilarity of turbulent transport of momentum and scalars in the unstable atmospheric surface layer", *Boundary-Layer Meteorol.*, Vol. 140 No. 2, pp. 243–262, (2011).

A. Liberzon, R. Gurka and G. Hetsroni, "Vorticity characterization in a turbulent boundary layer using PIV and POD analysis", *Proc. 4th Intl. Symp. Particle Image Velocimetry* Gottingen, Germany, 2001.

A. Liberzon, R. Gurka and G. Hetsroni, "XPIV multi-plane stereoscopic particle image velocimetry", *Exp. Fluids*, Vol. 36, No. 2, pp. 355–362, (2004).

A. Liberzon, R. Gurka and G. Hetsroni, "Comparison between two and three-dimensional POD in a turbulent boundary layer using multi-plane stereoscopic PIV", *J. Phys.: Conf. Ser.*, Vol. 318, No. 2, p. 022010, (2011).

A. Liberzon, R. Gurka, I. Tiselj and G. Hetsroni, "Spatial characterization of the numerically simulated vorticity fields", *Theoret. Comput. Fluid Dynam.*, Vol. 19, pp. 115–125, (2005).

A. Lozano-Durán and J. Jiménez, "Time-resolved evolution of coherent structures in turbulent channels: characterization of eddies and cascades", *J. Fluid Mech.*, Vol. 759, pp. 432–471, (2014).

J. L. Lumley, *"Stochastic Tools in Turbulence*, Applied Mathematics and Mechanics". Vol. 12 Academic Press, New York, (1970).

I. Marusic, "On the role of large-scale structures in wall turbulence", *Phys. Fluids*, Vol. 13, (2001).

K. G. McNaughton and Y. Brunet, "Townsend's hypothesis, coherent structures and monin–obukhov similarity", *Boundary-layer meteorology*, Vol. 102, No. 2, pp. 161–175, (2002).

I. Nezu and H. Nakagawa, *"Turbulence in Open-Channel Flows"*. IAHR Monograph Series. A. A. Balkema, Rotterdam, 1993.

R. Panton, *Self-Sustaining Mechanisms of Wall Turbulence*, Advances in Fluid Mechanics. Vol. 15, Computational Mechanics Publications, Southampton, UK, (1997).

S. Robinson, "Coherent motions in the turbulent boundary layer", *Annu. Rev. Fluid Mech.*, Vol. 23, pp. 601–639, (1991).

W. Schoppa and F. Hussain, "Coherent structure dynamics in near-wall turbulence", *Fluid Dynam. Res.*, Vol. 26, pp. 119–139, (2000a).

W. Schoppa and F. Hussain, "Coherent structure dynamics in near-wall turbulence", *Fluid Dynam. Res.*, Vol. 26, pp. 119–139, (2000b).

L. Sirovich, "Turbulence and the dynamics of coherent structures, Part I: Coherent structures", *Quar. Appl. Math.*, Vol. XLV, pp. 561–571, (1987).

C. Smith, "A synthesized model of the near-wall behavior in turbulent boundary layers", *Proc.8th Sympo Turbulence* Rolla, 1984.

C. Smith and J. Walker, "Turbulent wall-layer vortices", *Fluid Vortices*, S. Green (Ed.), Chapter VI, Kluwer Academic Publishers, Amsterdam, (pp. 235–283), (1995).

A. J. Smits, B. J. McKeon and I. Marusic, "High–Reynolds number wall turbulence", *Ann. Rev. Fluid Mech.*, Vol. 43, pp. 353–375, (2011).

Z. J. Taylor, R. Gurka, G. A. Kopp and A. Liberzon, "Long-duration time-resolved PIV to study unsteady aerodynamics", *IEEE Trans. Instrum. Measure.*, Vol. 59 No. 12, pp. 3262–3269, (2010).

T. Theodorsen, "Mechanism of turbulence", *Proc. 2nd Midwestern Conference on Fluid Mechanics*, 1952.

A. Tsinober, "Vortex stretching versus production of strain/dissipation", *Turbulence Structure and Vortex Dynamics*, J. Hunt and J. Vassilicos (Eds.), Cambridge, (pp. 164–191), (2000).

C. Willert, "Stereoscopic digital particle image velocimeter for application in wind tunnel flows", *Meas. Sci. Tech.*, Vol. 8, pp. 1465–1479, (1997).

Q. Zhong, Q. Chen, H. Wang, D. Li and X. Wang, "Statistical analysis of turbulent super-streamwise vortices based on observations of streaky structures near the free surface in the smooth open channel flow", *Water Resource. Res.*, Vol. 52 No. 5, pp. 3563–3578, (2016).

J. Zhou, R. Adrian, S. Balachandar and M. Kendall, "Mechanisms for generating coherent packets of hairpin vortices in channel flow", *J. Fluid Mech.*, Vol. 387, pp. 353–396, (1999).

Penetration of Turbulent Temperature Fluctuations into the Heated Wall

Iztok Tiselj

*Reactor Engineering Division "Jožef Stefan" Institute,
Jamova 39, SI-1000 Ljubljana, Slovenia*
iztok.tiselj@ijs.si

Conjugate heat transfer studies of heated wall being cooled by the turbulent flow are today considered as key benchmarks for thermal fatigue models being used in nuclear engineering. Remarkable contribution to the field had been made by Professor Gad Hetsroni and his collaborators through experimental studies and simulations performed between 1999 and 2004. The author of the present paper, a postdoctoral student of Professor Hetsroni in 1999, was responsible for computational studies of the phenomena. A review of these studies and of the research being done in the field by the author after 2004 is presented in the present paper. Direct Numerical Simulations (DNS) of the fully developed velocity and temperature fields in the turbulent channel flow coupled with the unsteady conduction in the heated walls were carried out. Simulations were performed with passive scalar approximations at Prandtl numbers of water and air. In the recent years, similar simulations were performed at low Prandtl number 0.01, which roughly corresponds to the Prandtl number of liquid sodium. Conjugate heat transfer simulations point to a relatively intensive penetration of turbulent temperature fluctuations into the heated wall for liquid sodium–steel system, which is relevant for fast reactor studies. Some standard statistical quantities like mean temperatures, profiles of the root-mean-square (RMS) temperature fluctuations for various thermal properties, and various wall thicknesses were obtained from these simulations. An interesting feature of the low Prandtl number simulations are large-scale turbulent structures, which are playing a much more important role at low Prandtl numbers than at the moderate Prandtl numbers of around one and required very large computational domains to capture the proper temperature fluctuation profiles. Recent efforts are focused on DNS in more complex geometries and on studies of wall-resolved LES models for conjugate heat transfer.

1. Introduction

Direct Numerical Simulation (DNS) of turbulent flows has emerged as an important research tool of turbulent heat transfer in the past three decades (Moin and Mahesh, 1999; Kasagi and Iida, 1999). Particular attention was focused on the DNS of the near-wall flows as they reveal the basic mechanisms of the convective heat transfer between the fluid and the solid wall. In the past decade, this topic is becoming important in the field of thermal fatigue problems (Brillant *et al.*, 2006; Aulery *et al.*, 2012).

The first DNSs of heat transfer in the channel flow geometry were made at low Reynolds numbers and at Prandtl numbers of around one by Kim and Moin (1989) and Kasagi *et al.* (1992). Later, Kawamura *et al.* (1998), Na and Hanratty (2000) performed DNS of the turbulent channel flow at Prandtl numbers up to 10.

The above-mentioned simulations used non-fluctuating temperature boundary condition for the dimensionless temperature. This assumption corresponds to the physical configuration where the fluid with negligible density ρ, specific heat capacity c_p, thermal conductivity λ is heated by a thick wall with high density, high heat capacity, and high thermal conductivity/diffusivity (thermal activity ratio $K = \sqrt{\rho_f c_{pf} \lambda_f / \rho_w c_{pw} \lambda_w} \to 0$).

Another example of thermal boundary condition is ideal fluctuating temperature boundary condition ($K = \infty$ or $\lambda_f \gg \lambda_w$). Comparison of fluctuating and non-fluctuating temperature boundary condition has been given by a group of Hetsroni (Tiselj, *et al.*, 2001a) who performed DNS of the turbulent flume flow with both ideal thermal boundary conditions and analyzed the differences at $\mathrm{Pr} = 1$ and $\mathrm{Pr} = 5.4$. They have shown that these two boundary conditions represent the two limiting cases of conjugate heat transfer. Figure 1, with results taken from the simulations in Tiselj, *et al.* (2001a), shows the differences between the two walls heated with equal constant volumetric heat source and cooled with the identical turbulent flow at the same Reynolds and Prandtl number. Passive scalar approximation is used in the simulation. Temperature field at the wall for the non-fluctuating boundary condition imposes constant wall temperature, thus a temperature field slightly above the wall is shown. When fluctuating temperature boundary condition is imposed at the wall, measurable temperature fluctuations appear at the solid–fluid contact plane. It is nevertheless interesting to note, that the heat transfer coefficients remain practically unchanged with the modification of the boundary condition (Tiselj, *et al.*, 2004).

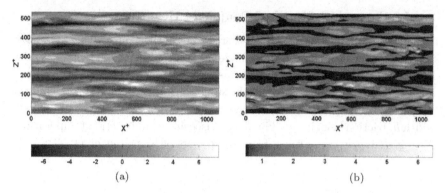

Fig. 1   Dimensionless temperature field at the heated wall ($Re_\tau = 180$, $Pr = 1$). (a) and (b) Fluctuating (at the wall; $y^+ = 0$) and non-fluctuating (just above the wall; $y^+ = 1.6$) wall temperature boundary condition, respectively.

To reveal all the details of the heat transfer along the wall with a given thickness and material properties ($\rho_w, c_{pw}, \lambda_w$), a conjugate turbulent heat transfer problem taking into account unsteady heat conduction inside the wall has to be solved. Conjugate heat transfer near the flat wall was studied by Polyakov (1974), Khabakhpasheva (1986), Sinai (1987), Kasagi *et al.* (1989), and Sommer (1994), while Tiselj *et al.* (2001b) in collaboration with the group of Hetsroni performed the first DNS study of the phenomena. Kasagi *et al.* (1989) have shown that in the experiments performed with air, the wall temperature fluctuations are usually negligible. The experiments with water are mostly performed at almost non-fluctuating temperature boundary condition when thick metal plates are used as heaters (Khabakh-pasheva, 1986). However, in specially designed experiments with very thin metal walls (foils), an approximation of fluctuating temperature boundary condition can be achieved with water.

Accurate measurements of the wall temperature fluctuations with nearly fluctuating temperature boundary condition were performed also by Mosyak *et al.* (2001) with water as a working fluid. Turbulent water flume in their experiment was heated with a very thin metal foil. Foil temperature was measured with the infra-red camera, and the measurements were in agreement with the results shown in Fig. 1(a).

For liquid sodium and stainless steel wall, the boundary condition falls somewhere in-between the fluctuating and non-fluctuating type. In the recent years, these materials have gained more attention due to the renewed interest in the sodium cooled nuclear reactors.

2. Mathematical Model of Conjugate Heat Transfer

The most frequently used geometry for studies of the near-wall heat transfer is turbulent channel flow. Computational setup of developed channel flow is shown in Fig. 2. Both walls are assumed to have the same thickness, identical material properties, and the same constant volumetric heat source.

The governing equations of the fluid, normalized with channel half width h, friction velocity u_τ, kinematic viscosity ν, and friction temperature $T_\tau = q_w/(u_\tau \rho_f c_{pf})$ are taken from the papers of Kasagi *et al.* (1992):

$$\nabla \cdot \vec{u}^+ = 0, \tag{1}$$

$$\frac{\partial \vec{u}^+}{\partial t} = -\nabla \cdot (\vec{u}^+ \vec{u}^+) + \frac{1}{\mathrm{Re}_\tau} \nabla^2 \vec{u}^+ - \nabla p + \vec{1}_x, \tag{2}$$

$$\frac{\partial \theta^+}{\partial t} = -\nabla \cdot (\vec{u}^+ \theta^+) + \frac{1}{\mathrm{Re}_\tau \cdot \mathrm{Pr}} \nabla^2 \theta^+ + \frac{u_x^+}{u_B^+}. \tag{3}$$

As seen from Eqs. (1)–(3), temperature is assumed to be passive scalar, which means that a single velocity field can be used for simulations with several temperature fields with different Prandtl numbers or with different material properties of the solid wall.

Dimensionless equation of heat conduction in the wall with internal heating is

$$\frac{\partial \theta^+}{\partial t} = \frac{1}{G \cdot \mathrm{Re}_\tau \cdot \mathrm{Pr}} \nabla^2 \theta^+ - \frac{K}{d^+ \sqrt{G}}, \tag{4}$$

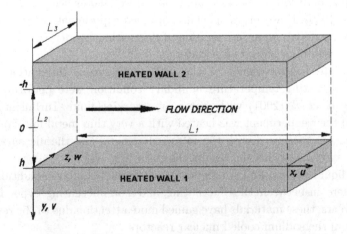

Fig. 2   Computational domain.

where $-K/(d^+\sqrt{G})$ is dimensionless internal heat source, $G = \alpha_f/\alpha_w$ and $K = \sqrt{(\rho_f c_{pf} \lambda_f)/(\rho_w c_{pw} \lambda_w)}$ — thermal activity ratio. This form of the dimensionless equation was used by Monod *et al.* (2012), who have pointed to the incompleteness of the conjugate heat transfer equations used by Kasagi *et al.* (1989) and Tiselj *et al.* (2001b). They have shown that variable thermal activity ratio $K = \sqrt{\rho_f c_{pf} \lambda_f / \rho_w c_{pw} \lambda_w}$ used in the previous studies should be combined with variable ratio of thermal diffusivities $G = \alpha_f/\alpha_w$, while the previous studies were performed for variable K and constant value of $G = 1$.

Boundary conditions for temperature and heat flux at the solid–fluid interface are:

$$\theta_f^+ = \theta_w^+, \quad K\sqrt{G}\frac{\partial\theta_f^+}{\partial y^+} = \frac{\partial\theta_w^+}{\partial y^+}. \tag{5}$$

Mean dimensionless temperature at both fluid–solid interfaces is fixed to zero: $\langle\theta^+(y = \pm 1)\rangle_{x,z,t} = 0$. External boundaries are adiabatic.

Material properties for selected fluid–solid combinations are listed in Table 1. The thermal activity ratio K is the most important for rule-of-the-thumb estimation of the thermal boundary condition type: very low K for air/metal means practically non-fluctuating temperature BC. For water/steel combination, the thermal BC is still close to non-fluctuating BC; however, it may approach to fluctuating BC if the metal wall thickness is small. For liquid sodium/steel combination with $K = 1$, the thermal boundary condition is roughly in-between the fluctuating and non-fluctuating temperature BC.

While all DNS simulations assume constant material properties, it is well known that these properties vary with temperature. However, the main purpose of DNS simulations is to produce a database that is useful for

Table 1 Material properties of relevant fluid–solid systems for conjugate heat transfer analyses.

	K	G	Pr
Air/aluminum	$3*10^{-4}$	0.3	0.7
Air/iron	$8*10^{-4}$	5	0.7
Water/aluminum	0.07	0.002	7
Water/steel (300 K)	0.2	0.01	7
Water/steel (600 K)	0.2	0.04	0.9
Liquid sodium/steel	1	10	0.005

verification of LES (Bricteux *et al.*, 2012) and RANS (Craft *et al.*, 2010) models of turbulent heat transfer at higher Reynolds number and in realistic geometries, where DNS cannot be performed.

3. Numerical Procedures

The first turbulent heat transfer code used by Tiselj and Hetsroni was based on pseudo-spectral scheme using Fourier series in x and z directions and Chebyshev polynomials in wall-normal y direction (Tiselj *et al.* 2001a). Second-order accurate time differencing (Crank–Nicholson scheme for diffusive terms and Adams–Bashfort scheme for other terms) is used with maximum CFL numbers of ~0.1. Aliasing error is removed with computation of the nonlinear terms on 1.5 times finer grid in each direction.

The code was verified on various occasions with other DNS data bases by (Kawamura *et al.*, 1998; Moser *et al.*, 1999) who performed DNS with and without heat transfer at various Reynolds and Prandtl numbers.

In 2001, the code was upgraded with the heat conduction equations in both walls that are solved with a mixture of spectral and finite difference method and are coupled with the fluid energy equation. In the latest simulations, fully spectral approach is used also in the solid domain (Tiselj *et al.*, 2013). The code allows solving several energy equations with different boundary conditions in parallel with a single velocity field solution. This approach reduces the CPU time but increases the required physical memory of the computer.

Each conjugate DNS computation was performed in two steps. The first step was computation with two temperature fields with ideal thermal boundary conditions: fluctuating temperature boundary condition ($\lambda_f \gg \lambda_w$) and non-fluctuating temperature BC ($\lambda_f \ll \lambda_w$). Statistical sampling started after the fully developed turbulent flow was established. The second step was conjugate heat transfer simulations, where a large number of temperature fields were simulated for various values of thermal activity ratio K, thermal diffusivity ratio G and various wall thicknesses, with slightly shorter averaging time.

Alternatives to spectral schemes are sought in order to extend the simulations to more complex geometries. Thus, some of these results have been recently reproduced with finite difference/volume schemes (Flageul *et al.*, 2015) and spectral element codes (Oder *et al.*, 2015).

Computational wall-times were (not surprisingly) the same from the first simulations in 1999, which were performed with ~50,000 time steps on a grid of 0.2 million spectral modes, to the latest simulations in 2014 with

~800,000 time steps on a grid of 80 million spectral modes. In all cases, the longest simulations took up to a few months.

4. DNS Database

Overview of the turbulent heat transfer simulations performed in the near-wall flow is collected in Table 2. Some of the first simulations were performed in the flume geometry with a single infinite wall at one side and a flat free surface on the other side of the fluid layer, while most of them were performed in the channel geometry.

The fourth column of Table 2 specifies thermal boundary conditions used in the *simulations*; simulations denoted with F and NF were performed only with the two limiting thermal boundary conditions: fluctuating and non-fluctuating temperature BC. Simulations denoted with C in the fourth column were true conjugate heat transfer simulations.

DNS at maximum Re_τ up to 590 were performed at various Prandtl numbers. An exception are high Prandtl number simulations considered in the paper by Bergant and Tiselj (2007): thermal fields in high Pr number DNS would require resolution, which is finer than the velocity field resolution. Namely, the Batchelor scale (smallest spatial scale of the temperature

Table 2   Review of the DNS database. F — Fluctuating temperature BC, NF — Non-fluctuating temperature BC, C — conjugate heat transfer simulation.

Simulation type, geometry	Re_τ	Pr	Thermal boundary conditions	K	G	solid wall thickness d^+	Ref.
DNS flume	170	1, 5.4	F vs. NF	/	/	.	9
DNS channel	150	0.7, 7	C	0.1–100	1	0.5–20	10
DNS flume	170	1, 54	F vs. NF	/	/	/	25
DNS channel and flume	Channel 150 Flume 424	Channel 0.025, 0.7 Flume: 1, 5.4	F vs. NF	/	/	/	11
DNS-LES channel	150, 395	25, 100, 200	NF	/	/	/	26
DNS channel	150, 395, 590	0.01	C	0.01– 100	0.001– 1000	0.1– 1000	27
DNS periodic channel	180, (395)	1, (1, 10)	C	0.01– 100	0.001– 100	0.001– 540	22
DNS channel	180	0.01, 1	F vs. NF	/	/	/	30
DNS channel spec. elements	180	0.01	C	1	1	180	24

field) is $Pr^{1/2}$ smaller than the Kolmogorov scale. The main result of the simulations at high Pr number was the finding that temperature scales smaller than the Kolmogorov scale do not have a significant effect on the mean temperature and on the temperature fluctuation profiles. Nevertheless, the simulations at high Pr numbers by Bergant and Tiselj (2007) are denoted as DNS-LES; DNS for velocity and LES for temperature field.

Some recent simulations (Tiselj *et al.*, 2012; Oder *et al.*, 2015) were performed with very low Prandtl numbers as benchmarks for the codes that will be used for heat transfer analyses in liquid sodium fast reactors. The latest publication (Oder *et al.*, 2015) summarizes results of channel flow conjugate heat transfer simulations with spectral element code NEK5000.

Slightly different configuration was studied by Tiselj *et al.* (2013): double-sided cooling of the slab was analyzed, where temperature fluctuations penetrate the heater from both sides. Due to the non-coherent turbulent flows on each side of the slab, thermal fluctuations in a thin slab are actually lower than in the same slab that is cooled by the same turbulent flow from a single side and is insulated on the other side.

Additional analysis of turbulent heat transfer with fluctuating and non-fluctuating temperature boundary conditions was done by Tiselj (2014) at low Reynolds ($Re_\tau = 180$) and low Prandtl number $Pr = 0.01$. The largest computational domain in this study was six times longer and six times wider than the "standard" computational domain in our previous DNS studies or in the work of Kim and Moin (1989). Comparison of the standard and large computational domains has shown the velocity field statistics (mean velocity, root-mean-square, RMS fluctuations, and turbulent Reynolds stresses) that are within 1%–2%. Similar agreement was observed for $Pr = 1$ temperature fields and also for the mean temperature profiles at $Pr = 0.01$. These differences were attributed to the statistical uncertainties of the DNS. However, second-order moments, i.e. RMS temperature fluctuations of standard and large computational domains at $Pr = 0.01$ have shown significant differences of up to 20%. Stronger temperature fluctuations in the very large domain confirmed the existence of the large-scale structures. Their influence is invisible in the main velocity field statistics or in the statistics of the temperature fields at Prandtl numbers around unity. However, they play a visible role in the temperature fluctuations at low Prandtl number, where high temperature diffusivity effectively smears the small-scale structures in the thermal field and enhances the relative contribution of large-scales. This effect is visible (Tiselj, 2014) also at $Re_\tau = 395$ but remains to be confirmed at higher Reynolds numbers.

5. Results

Selected results from the simulations in Table 2 are summarized in this section. Mean temperature profiles are the most important information for development of heat transfer coefficient correlations. Figures 3 and 4 show several mean temperature profiles obtained at various Reynolds and Prandtl numbers. Profiles at $Pr = 1$ and $Pr = 0.01$ are given for both types of ideal boundary conditions, and high Pr results are shown for non-fluctuating

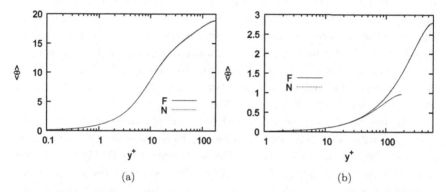

Fig. 3   Mean temperature profiles at various Prandtl and Reynolds numbers. (a) $Re_\tau = 180$, $Pr = 1$ (Tiselj, 2014), (b) $Re_\tau = 180$, 590, $Pr = 0.01$ (Tiselj, 2004; Tiselj and Cizelj, 2012).

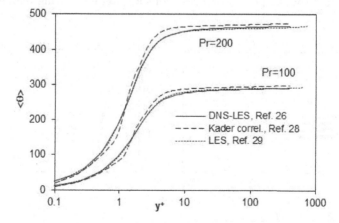

Fig. 4   Mean temperature profiles at $Re_\tau = 395$ and $Pr = 100$ and $Pr=200$, DNS-LES results (Bergant and Tiselj, 2007; Kader, 1981; Calmet and Magnaudet, 1997).

temperature BC. The main result concerning the conjugate heat transfer and the mean temperature profiles is actually the fact that the type of thermal boundary condition, i.e. fluctuating or non-fluctuating temperature BC, does not have a significant influence on the mean temperature profile and on the heat transfer coefficient. The study performed by Tiselj *et al.* (2004) has shown that the heat transfer coefficient for the fluctuating temperature BC is up to 1% higher (typically around 0.5% higher) than for the non-fluctuating temperature BC. However, this difference is difficult to measure even in the DNS simulations, since it is roughly comparable with statistical uncertainty of the DNS results. Results of DNSs shown in Fig. 3 were obtained after 2004 and confirm the conclusions in Tiselj *et al.* (2004): profiles with fluctuating temperature BC are consistently around ∼0.5% higher than the non-fluctuating temperature BC at the same Reynolds and Prandtl numbers: 0.6% (Pr = 1), 0.3% (Pr = 0.01, $Re_\tau = 180$), 0.8% (Pr = 0.01, $Re_\tau = 590$).

The key results of the DNS with variable thermal boundary conditions and with conjugate heat transfer are thus temperature fluctuations profiles. Figure 5 shows several profiles of temperature RMS fluctuations from various simulations in Table 2. Only the results for the two limiting boundary conditions are shown in Fig. 5: the non-fluctuating temperature BC profiles approaches zero at the wall, while the fluctuating temperature BC approaches a constant value. Rough value of the temperature fluctuations at the wall for fluctuating temperature BC is comparable with maximum of the temperature fluctuations for non-fluctuating temperature

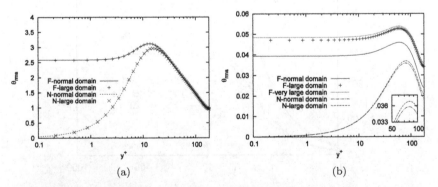

Fig. 5 Temperature RMS fluctuations profiles for fluctuating (F) and non-fluctuating (N) temperature BC at $Re_\tau = 180$. (a) Pr=1, (b) Pr=0.01 (data and figure from Tiselj, 2014).

BC that is typically measured at the distance $\sim 20/\text{Pr}^{1/2}$ wall units away from the wall. Precise value of the wall fluctuations can be predicted only with DNS.

An interesting feature regarding the near-wall temperature fluctuations is seen in Fig. 5, right, where the near-wall fluctuations at low Prandtl number 0.01 depend on the size of the computational domain. The paper by Tiselj (2014) explains the difference, which stems from the very large turbulent structures in the near-wall flow. Contribution of these scales in the energy spectra of the velocity fluctuations is negligible; however, it is not negligible in the low Prandtl number temperature field, where very high thermal diffusivity efficiently eliminates the small-scale structures and enhances the role and visibility of the large structures.

The true conjugate heat transfer results are shown in Figs. 6 and 7. Figure 6 shows results at $\text{Re}_\tau = 150$, $\text{Pr} = 7$, $G = 1$ from the first simulations of Tiselj *et al.* (2001b). Figure 6 (left) shows temperature fluctuations inside the solid wall of thickness $d^+ = 20$ and Fig. 6 (right) shows fluctuations inside the fluid for variable thermal activity ratios K. The $K = 1$ profile is roughly in the middle between the ideal fluctuating and ideal non-fluctuating profiles in the near-wall region, and it turns out that this is true also for Reynolds numbers examined at liquid sodium Prandtl number $\text{Pr}=0.01$ and moderate values of G (Tiselj and Cizelj, 2012). Some of these results are shown in Fig. 7 at $\text{Re}_\tau = 395$, $\text{Pr} = 0.01$, $G = 1$.

The most convenient way to determine the temperature fluctuations at the fluid–solid plane is to use charts shown in Fig. 8, which are valid for

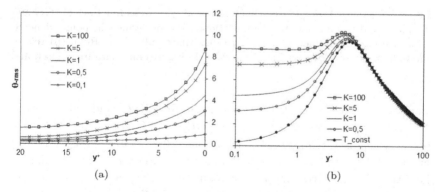

Fig. 6 Temperature RMS fluctuations in the solid (a) and in the fluid domain (b) at $\text{Re}_\tau = 150$, $\text{Pr}=7$, $d^+ = 20$, variable K and $G = 1$ (data from Tiselj *et al.*, 2001b).

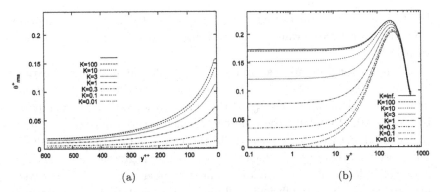

(a)   (b)

Fig. 7   RMS temperature fluctuations profiles at Pr=0.01 and $Re_\tau = 395$ for different thermal activity ratios K, constant thermal diffusivity ratio $G = 1$, and thick wall $d^+ = 395$. (a) Fluctuations inside the wall; (b) fluctuations in the fluid (data from Tiselj and Cizelj, 2012).

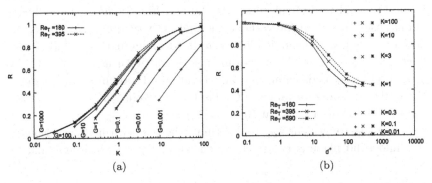

(a)   (b)

Fig. 8   Normalized temperature fluctuations at $y^+ = 0$ as functions of K, G and d^+ for Prandtl number 0.01. Computations were performed in points denoted with symbols. (a) influence of K and G for thick wall ($d^+ = Re_\tau$). (b) role of the wall thickness d at variable K and $G = 1$ (Figures and data from Tiselj and Cizelj, 2012).

Pr=0.01. One needs to evaluate parameters Re_τ, K, G, and wall thickness d^+, which depend on the material properties and fluid velocity. The value of the normalized temperature fluctuations ($R(y^+ = 0) = \theta/\theta_{\text{fluctuatingBC}}$) at the solid–fluid plane $y^+ = 0$ determines the importance of the conjugate heat transfer and gives an estimate about the maximum temperature fluctuations in the solid. Normalization factors — RMS values $\theta_{\text{fluctuatingBC}}$ at the wall for fluctuating temperature BC are: 0.040, 0.10, 0.17, at $Re_\tau = 180$, 395, and 590, respectively. It turns out (Tiselj and Cizelj, 2012) that the normalized temperature fluctuations show only weak dependence on the Reynolds number. Similar charts for $Pr = 1$ and $Pr = 7$ (but only for $G = 1$) are available in Tiselj *et al.* (2001b).

Other turbulent quantities, like turbulent heat fluxes, spectra of the temperature fluctuations, budget terms of the turbulent quantities, etc., also exhibit differences in the near wall region. The reader is referred to the references in Table 2 for some of these details.

Our recent efforts are focused on spectral element method and open source code NEK5000, which will allow DNS in slightly more complex geometries than the pure spectral method. Some of the first results are collected in the paper by Oder *et al.* (2015). Our second research direction is development and validation of the wall-resolved LES models for conjugate heat transfer. These are based on finite volume methods and open source Code-Saturne (Flageul *et al.*, 2015). The key benchmarks for validation of these simulations can be found in the database of Hestroni and his collaborators.

6. Conclusions

The legacy of professor Hestroni in the field of multiphase flows is invaluable. Nevertheless, he left important contributions also in some other fields of single-phase flow and heat transfer. One of these fields is the study of the turbulent heat transfer near the flat wall. Measurements and simulations of turbulent heat transfer that were done under his leadership clarified the phenomena of the thermal turbulent fluctuations that penetrate the solid wall of the heater. His work in this field remains highly visible, and the author of this paper is proud to be a part of this story.

From: Gad Hetsroni [mailto:hetsroni@techunix.technion.ac.il]
Sent: Wednesday, November 26, 2003 3:12 PM
To: Tiselj, Iztok
Subject: RE: IJHMT-JWR/AB 03 045

iztok
you want something done-give it to a busy person.
gh

Copyright

Fully Developed Turbulent Flume Flow. Phys. Fluids 13 (4) (2001), with the permission of AIP Publishing.

Fig. 5 reprinted from I. Tiselj, Tracking of Large-Scale Structures in Turbulent Channel with DNS of Low Prandtl Number Passive Scalar. Phys. Fluids 26 (12) (2014), with the permission of AIP Publishing.

Figs. 7, 8, reprinted from Nuclear Engineering and Design 253, (2012), I. Tiselj, L. Cizelj, DNS of turbulent channel flow with conjugate heat transfer at Prandtl number 0.01. Copyright (2012) with permission of Elsevier.

References

1. P. Moin and K. Mahesh, "Direct numerical simulation: A tool in turbulence research", *Ann. Rev. Fluid Mech.*, Vol. 30, pp. 539–578, (1999).
2. N. Kasagi and O. Iida, "Progress in direct numerical simulation of turbulent heat transfer", *Proceedings of the 5th ASME/JSME Joint Thermal Engineering Conference*, American Society of Mechanical Engineers, San Diego, USA, 1999.
3. G. Brillant, S. Husson and F. Bataille, "Subgrid-scale diffusivity: Wall behaviour and dynamic methods", *J. Appl. Mech., Trans ASME*, Vol. 73, No. 3, pp. 360–367, (2006).
4. F. Aulery, A. Toutant, G. Brillant, R. Monod and F. Bataille, "Numerical simulations of sodium mixing in a T-junction", *Appl. Therm. Eng.*, Vol. 37, pp. 38–43, (2012).
5. J. Kim and P. Moin, "Transport of Passive Scalars in a Turbulent Channel Flow", *Turbulent Shear Flows VI*, Springer-Verlag, Berlin, 1989, pp. 85.
6. N. Kasagi, Y. Tomita and A. Kuroda, "Direct numerical simulation of passive scalar field in a turbulent channel flow", *J. Heat Tr. — Tran. ASME*, Vol. 114, pp. 598–606, (1992).
7. H. Kawamura, K. Ohsaka, H. Abe and K. Yamamoto, "DNS of turbulent heat transfer in channel flow with low to medium-high Prandtl number fluid", *Int. J. Heat Fluid Flow*, vol. 19 pp. 482–491, (1998).
8. Y. Na and T. J. Hanratty, "Limiting behavior of turbulent scalar transport close to a wall", *Int. J. Heat Mass Tran.*, Vol. 43, pp. 1749–1758, (2000).
9. I. Tiselj, E. Pogrebnyak, Changfeng Li, A. Mosyak and G. Hetsroni, "Effect of wall boundary condition on scalar transfer in a fully developed turbulent flume flow", *Phys. Fluids*, Vol. 13, No. 4, pp. 1028–1039, (2001a).
10. I. Tiselj, R. Bergant, B. Mavko, I. Bajsic and G. Hetsroni, "DNS of turbulent heat transfer in channel flow with heat conduction in the solid wall", *J. Heat Trans. — Tran. ASME*, Vol. 123, pp. 849–857, (2001b).
11. I. Tiselj, A. Horvat, B. Mavko, E. Pogrebnyak, A. Mosyak and G. Hetsroni, "Wall properties and heat transfer in near wall turbulent flow", *Num. Heat Tran. A*, Vol. 46, pp. 717–729, (2004).
12. A. F. Polyakov, "Wall effect of temperature fluctuations in the viscous sublayer", *Teplofizika Vysokih Temperatur*, Vol. 12, pp. 328–337, (1974).

13. Y. M. Khabakhpasheva, "Experimental investigation of turbulent momentum and heat transfer in the proximity of the wall", *Proc. 8th Int. Heat Transfer Conference*, San Francisco, USA, 1986.

14. Y. L. Sinai, "A wall function for the temperature variance in turbulent flow adjacent to a diabatic wall", *J. Heat Tran. — T. ASME*, Vol. 109, pp. 861–865, (1987).

15. N. Kasagi, A. Kuroda and M. Hirata, "Numerical investigation of near-wall turbulent heat transfer taking into account the unsteady heat conduction in the solid wall", *J. Heat Tran. — T. ASME*, Vol. 111, pp. 385–392, (1989).

16. R. Monod, G. Brillant, A. Toutant and F. Bataille, "Large eddy simulations of a turbulent periodic channel with conjugate heat transfer at low Prandtl number", *J. Phy.: Conf. Ser.*, Vol. 395, No. 1, p. 012026, (2012).

17. T. P. Sommer, R. M. C. So and H. S. Zhang, "Heat transfer modeling and the assumption of zero wall temperature fluctuations", *J. Heat Tran. — T. ASME*, Vol. 116, pp. 855–863, (1994).

18. A. Mosyak, E. Pogrebnyak and G. Hetsroni, "Effect of constant heat flux boundary condition on wall temperature fluctuations", *J. Heat Tran. — T. ASME*, Vol. 123, pp. 213–218, (2001).

19. L. Bricteux, M. Duponcheel, G. Winckelmans, I. Tiselj and Y. Bartosiewicz, "Direct and large eddy simulation of turbulent heat transfer at very low Prandtl number: Application to lead–bismuth flows", *Nuc. Eng. Design*, Vol. 246, pp. 91–97, (2012).

20. T. J. Craft, H. Iacovides and S. Uapipatanakul, "Towards the development of RANS models for conjugate heat transfer", *J. Turbulence*, Vol. 11, pp. 1–16, (2010).

21. D. R. Moser, J. Kim and N. N. Mansour, "Direct numerical simulation of turbulent channel flow up to $Re_\tau = 590$", *Phys. Fluids*, Vol. 11, no. 4, pp. 943–945, (1999).

22. I. Tiselj, J. Oder and L. Cizelj, "Double-sided cooling of heated slab: Conjugate heat transfer DNS", *Int. J. Heat Mass Tran.*, Vol. 66, pp. 781–790, (2013).

23. C. Flageul, S. Benhamadouche, E. Lamballais and D. Laurence, "DNS of turbulent channel flow with conjugate heat transfer: Effect of thermal boundary conditions on the second moments and budgets", *Int. J. Heat Fluid Flow*, Vol. 55, pp. 34–44, (2015).

24. J. Oder, J. Urankar and I. Tiselj, "Spectral element direct numerical simulation of heat transfer in turbulent channel sodium flow", *24th International Conference Nuclear Energy for New Europe*, Portorož, Slovenia, 2015.

25. G. Hetsroni, I. Tiselj, R. Bergant, A. Mosyak and E. Pogrebnyak, "Convective velocity of temperature fluctuations in a turbulent flume", *J. Heat Tran. — T. ASME*, Vol. 126, No. 5, pp. 843–848, (2004).

26. R. Bergant and I. Tiselj, "Near-wall passive scalar transport at high Prandtl numbers", *Phys. Fluids*, Vol. 19, No. 6, p. 065105, (2007).

27. I. Tiselj and L. Cizelj, "DNS of turbulent channel flow with conjugate heat transfer at Prandtl number 0.01", *Nuc. Eng. Design*, Vol. 253, pp. 153–160, (2012).

28. B. A. Kader, "Temperature and concentration profiles in fully turbulent boundary layers", *Int. J. Heat Mass Tran.*, Vol. 24, p. 1541, (1981).

29. I. Calmet and J. Magnaudet, "Large-eddy simulation of high-Schmidt number mass transfer in a turbulent channel flow", *Phys. Fluids*, Vol. 9, p. 438, (1997).

30. I. Tiselj, "Tracking of large-scale structures in turbulent channel with DNS of low Prandtl number passive scalar", *Phys. Fluids* Vol. 26, No. 12, p. 125111, (2014).

Experimental Investigation of the Interaction Between a Stationary Rigid Sphere and a Turbulent Boundary Layer

René van Hout[*,‡], Jerke Eisma[†], Edwin Overmars[†],
Gerrit E. Elsinga[†] and Jerry Westerweel[†,§]

[*] Technion — Israel Institute of Technology
Technion City, Haifa, Israel
[†] TU-Delft, 3ME Aero- and
Hydrodynamics Laboratory Leeghwaterstraat 21,
2628 CA, Delft, The Netherlands
[‡] rene@technion.ac.il
[§] J.Westerweel@tudelft.nl

Time-resolved tomographic particle image velocimetry (PIV) measurements (acquisition rate 250 Hz) were performed in a turbulent boundary layer on the side wall of an open channel, water flow facility (cross section $60 \times 60 \, \text{cm}^2$, $W \times H$), 3.5 m downstream of the inlet at a bulk flow velocity of $U_b = 0.17 \, \text{m/s}$ ($\text{Re}_b = U_b H / \nu = 97,679$, $\delta_{0.99} = 45.0 \, \text{mm}$, $\text{Re}_\theta = 752$). The measurement volume was a horizontal slab ($60 \times 15 \times 60 \, \text{mm}^3$, $l \times w \times h$) extending from the wall, 30 cm above the bottom. The setup comprised four high-speed ImagerPro HS cameras (2016×2016 pixels), a high-speed laser (Nd:YLF, Darwin Duo 80M, Quantronix), optics/prisms, and data acquisition/processing software (LaVision, DaVis 8.2). Data were acquired with and without a stationary held sphere that had a diameter, $D = 6 \, \text{mm}$ ($D^+ = 51$, "+" denotes inner wall scaling), and was positioned at $x_3 = 5.4$ and 37.6 mm ($x_3^+ = 43$ and 306) from the wall (measured from the sphere's center). Sphere Reynolds numbers based on D and the average streamwise velocity at the sphere's center were 692 and 959, respectively. The mean streamwise velocity profiles of the undisturbed boundary layer clearly exhibit a canonical shape. Introducing the sphere strongly affected log layer and buffer layer mean velocity and Reynolds stress profiles. Recovery to the undisturbed boundary layer characteristics is faster with the sphere positioned closest to the wall. When positioned at $h^+ = 306$, near-wall, uplifted, coherent vortical structures extend from the wall up to the sphere's wake with which they interact.

Keywords: Particle-laden flow; sphere; turbulent boundary layer.

1. Introduction

Particle–fluid interactions commonly occur in both industrial applications as well as in environmental settings. While in many cases two-way or four-way coupling often occurs, most research has been focused on one-way coupling. Here, the interaction between the wake flow formed behind a single, stationary held sphere with a fully developed turbulent wall boundary layer is investigated with the aim of resolving two-way coupling effects.

Many investigations, especially experimental visualizations and numerical simulations, have studied the wake flow behind a sphere in a uniform flow at moderate Reynolds numbers. It is well accepted that vortex shedding starts when $Re_D = U_\infty D/\nu$ exceeds 270 (Johnson and Patel, 1999; Sakamoto and Haniu, 1990). Initially, vortex shedding is highly organized and characterized by well-ordered sequences of entwined hairpin vortices with alternating direction of their "heads" (Johnson and Patel, 1999). As Re_D exceeds 800, the large vortical structures start to disintegrate into smaller ones as a result of shear layer instabilities (Sakamoto and Haniu, 1990) and the normalized vortex shedding frequency as characterized by the Strouhal number displays both a low and high frequency branch.

Few attempts have been made to measure the instantaneous 3D structure of the sphere wake, with a notable exception by Brücker (2001). He performed particle image velocimetry (PIV) measurements in a transverse plane positioned in the near-wake of a stationary held sphere from which he reconstructed the 3D wake structure assuming a constant convection velocity. He showed that well-ordered packets of shed vortices are regularly disturbed into a less well-ordered sequence after which the well-ordered packets re-emerge.

The effect of free stream turbulence on the sphere wake characteristics and drag force has been studied experimentally by Wu and Faeth (1994) and numerically by Bagchi and Balachandar (2003). The former showed that neither the onset of vortex shedding nor the vortex shedding frequency was affected by a free stream turbulence intensity of 4% in the range of $135 < Re_D < 1560$. They further showed that the wake flow was self-similar when taking into account enhanced turbulent viscosities. Direct numerical simulations (DNS) by Bagchi and Balachandar (2003) showed that the mean drag force could be well predicted by the standard drag relation without added mass and history effects. The accuracy of instantaneous drag prediction diminished with increasing D.

Zeng *et al.* (2008), using DNS, studied the effects of wall induced turbulence on the time dependent forces acting on a stationary held sphere in a turbulent channel flow. The sphere having a normalized diameter ranging from $3.5 < D^+ < 25$, (the superscript "+" denotes inner wall scaling) was located in the buffer layer and at the channel center. Sphere Reynolds numbers ranged from $42 < \mathrm{Re}_D < 295$ in the buffer layer and $\mathrm{Re}_D = 325$ and 455 at the channel center. They showed that the instantaneous forcing on the sphere could be correlated with the passage of near-wall coherent turbulence structures.

In the present study, we measured for the first time the dynamics of a 3D sphere wake interacting with a turbulent boundary layer. In order to achieve this we used temporally resolved tomographic particle image velocimetry (tomo-PIV). Here, results showing the effect of the sphere on the mean streamwise velocity and Reynolds stress profiles at different positions relative to the sphere will be presented. In addition, results on instantaneous near-wall structures and those in the sphere's wake are presented, as well as their possible interaction.

2. Description of Experiments

Experiments were performed in the water tunnel of the laboratory for Aero- and Hydrodynamics at Delft University of Technology. The tunnel had a cross section of $600 \times 600\,\mathrm{mm}^2$. Due to setup constraints, measurements were performed in the boundary layer that was formed on the vertical, channel side-wall made of transparent plexiglass. In order to force transition to turbulence, a zigzag strip (Elsinga and Westerweel, 2001) was placed 500 mm downstream of the test section's inlet (Fig. 1).

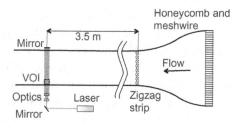

Fig. 1   Schematic of water channel and measurement position (top view).

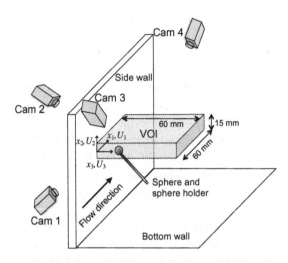

Fig. 2 Schematic of camera positions and VOI.

The tomo-PIV setup that is schematically shown in Figs. 1 and 2, comprised four high-speed ImagerPro HS cameras (2016×2016 pixels), a high-speed laser (Nd:YLF, Darwin Duo 80M, Quantronix), optics/prisms, and data acquisition/processing software (LaVision, DaVis 8.2). Here, x_i ($i = 1$, 2, 3) denote the streamwise, transverse and wall-normal coordinates; U_i and u_i denote the corresponding instantaneous and fluctuating velocities. An overbar indicates temporal averaging and angle brackets, $\langle v \rangle$, indicate spatial averaging.

Time resolved tomo-PIV measurements were performed 3.5 m downstream of the trip at a bulk flow velocity of $U_b = 0.17$ m/s, corresponding to a bulk Reynolds number, $\mathrm{Re}_b = U_b H / \nu = 97,679$, where $H = 60$ cm is the channel depth and ν is the kinematic viscosity. The volume of interest (VOI) consisted of a horizontal slab with dimensions of $60 \times 15 \times 60$ mm^3 ($l \times w \times h$). In order to make sure that the light intensity was sufficient and cameras benefitted from forward light scattering, a mirror was used to reflect the light back (Fig. 1). Additionally, this helped to diminish the shadow cast by the sphere. The VOI was located in the middle of the channel, 30 cm above the bottom wall. As flow tracers, near-neutrally buoyant hollow glass spheres (Sphericell, Potter's industries) were used. The friction velocity, $u_\tau = 0.0085$ m/s, was determined by a Clauser fit (von Kármán constant $\kappa = 0.41$ and intercept $B = 5.0$). A sphere with diameter, $D = 6.0$ mm, was held in place by a cylindrical rod ($d = 1.0$ mm) at two

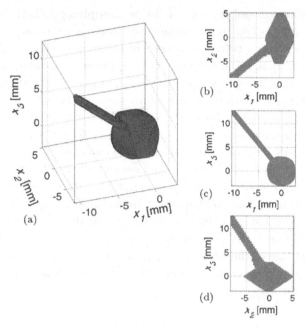

Fig. 3   3D reconstruction of (a) the sphere and holder (visual hull) and (b)–(d) its projections onto different planes.

different distances from the wall, i.e. $h = 5.4$ and 37.6 mm (measured from the sphere center to the wall), corresponding to $h^+ = 43$ and 306, respectively. Data sets were acquired at an acquisition frequency, $f_a = 250$ Hz, one set without a sphere and one for each sphere position, each set comprising 3139 instantaneous vector maps.

Based on projections of the sphere and holder in the four different camera views, a binary mask was constructed. These masks were imported back into the DaVis 8.2 software (LaVision) and were reconstructed using the "fast-MART" algorithm in the same way as particles in the VOI. The resulting visual hull (Adhikari and Longmire, 2012) is depicted in Fig. 3a at the same resolution as reconstructed vector maps (10 voxels in depth). Projections of the visual hull onto different planes are shown in Figs. 3b–d. Clearly, the sphere's visual hull resembles a double cone and only the projection in the $x_1 - x_3$ plane is nearly circular. Note that the exact shape of the visual hull depends on the camera setup and that the reconstructed visual hull would be more similar to the actual sphere when using additional cameras having different viewing angles. The visual hull was used to mask the reconstructed particle volume by multiplying each $x_1 - x_3$ plane at different depth (x_2) positions by the mask. Note that it is essential to mask the reconstructed particle volume since at the sphere position, ghost particles appeared upon reconstruction affecting the cross-correlation algorithm and leading to spurious vectors in the neighborhood of the sphere.

Reconstructed, masked particle volumes were processed using direct correlation of 3D particle volumes (DaVis 8.2, LaVision software) in multiple steps reducing the interrogation volume size from $96 \times 96 \times 96$ to $40 \times 40 \times 40$ voxels with 75% overlap in the last step. In between steps, the obtained vector maps were checked for outliers and smoothed by a $3 \times 3 \times 3$ Gaussian median filter. Vector maps were exported into Matlab where they were subsequently temporally and spatially smoothed using second-order polynomial regression over 7 points (Elsinga *et al.*, 2010). The size of the regression interval was of the order of the viscous time scale, $\nu/u_\tau^2 \approx 14\,\text{ms}$, and the smallest size of the near wall structures ($\sim$15 inner wall units) and is not expected to remove any relevant information on the near-wall turbulent flow structures.

3. Results

3.1. *Mean velocity and rms profiles without the sphere*

As an initial check of data quality, the processed vector maps were time averaged and spatially averaged across depth (the middle one-third of the VOI) and data were Reynolds decomposed. The obtained mean, streamwise velocity profile plotted in inner wall units is compared to available literature data in Fig. 4. The boundary layer thickness was $\delta_{0.99} = 45.0\,\text{mm}$, the displacement thickness, $\delta^* = 6.73\,\text{mm}$, momentum thickness, $\theta = 4.79\,\text{mm}$, and shape factor, $\delta^*/\theta = 1.41$.

The present data indicates that a log layer exists that extends up to $x_3^+ \approx 150$, of smaller extent than the data by DeGraaff and Eaton (2000) at higher Re_θ. The first two data points in the viscous sublayer are not well-resolved, but the present measurements collapse well with results by DeGraaff and Eaton (2000) in the buffer layer and log layer. A further comparison of the mean, streamwise component of the normal Reynolds stress, plotted in inner wall units together with available literature data is presented in Fig. 5. The present data are clearly inaccurate in the near-wall region for $x_3^+ < 10$ but do indicate a peak value at about $x_3^+ = 16$ as expected. Note that this peak value decreases with decreasing Re_θ as can be deduced from the data presented in Fig. 5 at decreasing Re_θ (DeGraaff and

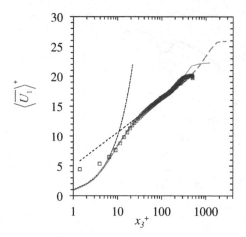

Fig. 4 Mean streamwise velocity profile plotted in inner wall coordinates. $\square$ Present data, $\text{Re}_\theta = 752$ (DeGraaff and Eaton, 2000), $\underline{}$ $\text{Re}_\theta = 1430$, $\cdots$ $\text{Re}_\theta = 5200$.

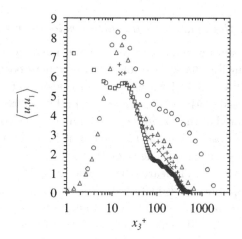

Fig. 5 Normal Reynolds stress profiles plotted in inner wall coordinates. $\square$ Present data, $\text{Re}_\theta = 752$; (DeGraaff and Eaton, 2000), $\triangle$ $\text{Re}_\theta = 1430$, $\bigcirc$ $\text{Re}_\theta = 5200$ (Erm and Joubert, 1991), $\times$ $\text{Re}_\theta = 697, +\,\text{Re}_\theta = 1003$.

Eaton, 2000; Erm and Joubert, 1991). Further, note that the peak value was attenuated by the regression that was applied and was 6% higher when the data were not temporally filtered. Reynolds wall-normal and shear stresses also compared well to available literature data (not shown) and behaved much better for $x_3^+ < 20$, reducing to near zero values as expected.

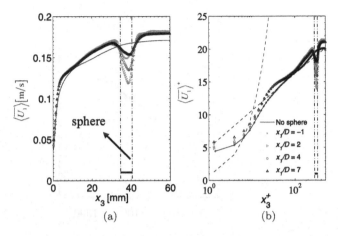

(a)   (b)

Fig. 6 Comparison between the mean, wall-normal streamwise velocity profiles at different positions relative to the sphere ($h^+ = 306$) and without sphere plotted (a) dimensional, and (b) in inner–inner scaling. Horizontal black bar indicates sphere position and dash-dot lines its extent.

3.2. *Effect of the sphere on the mean velocity profiles*

Introducing the stationary sphere affected the mean streamwise velocity profiles plotted in Fig. 6a with respect to the sphere positioned at $h^+ = 306$. In Secs. 3.2 and 3.3, data with the sphere present were temporally averaged and subsequently spatially averaged in the x_2 direction over a wall-normal slice centered at the sphere's center having a depth of $\Delta x_2 = 2.2\,\text{mm}$ ($\Delta x_2^+ = 17.5$). As observed in Fig. 6a, the flow is faster in the vicinity of the sphere as a result of blockage. The velocity deficit in the wake of the sphere slowly recovers but is still clearly discernible at $x_1/D = 7$. In addition, the shape of the profiles is also affected, which is further emphasized when plotting the mean profiles in inner wall units (Fig. 6b). Clearly, the log layer has nearly vanished, while profiles collapse for all x_1/D. Note that the effect of the sphere is already felt upstream at $x_1/D = -1$.

Since normalization by u_τ determined for the undisturbed turbulent boundary layer flow is questionable given the effect that the sphere has on the near-wall flow, outer wall scaling was applied through the boundary layer edge velocity, U_e, determined from the measured mean profiles at different x_1/D. Results are depicted in Fig. 7.

Normalization by U_e does not change the main conclusions that were drawn previously and Figures 6b and 7a look similar. A more detailed look at the changes in the log layer is provided in Fig. 7b which clearly indicates

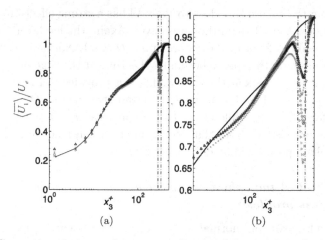

Fig. 7 Comparison between the mean, wall-normal streamwise velocity profiles at different positions relative to the sphere ($h^+ = 306$) and without sphere plotted in (a) outer–inner scaling and (b) the same as (a) at higher magnification. For legend see Fig. 6.

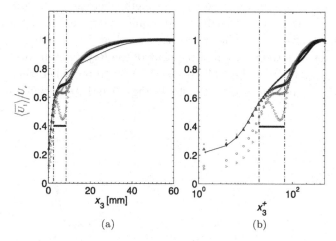

Fig. 8 Comparison between the mean, streamwise velocity profiles at different positions relative to the sphere ($h^+ = 43$) and without sphere plotted in (a) outer and (b) outer–inner scaling. Legend see Fig. 6.

that the log layer has been modified by the sphere's presence at all presented x_1/D.

Similar results for the mean streamwise velocity profiles normalized by U_e are plotted in Fig. 8 for the sphere positioned at $h^+ = 43$. As expected, introducing the sphere this close to the wall has a strong effect on both the near wall flow as well as on the outer layer. Again, the log layer has disappeared already in front of the sphere at $x_1/D = -1$. Normalized velocities are elevated above the sphere becoming the same at the edge of the boundary layer. A closer look at the near-wall flow is provided in Fig. 8b showing that at $x_1/D = -1$ and 7, normalized mean streamwise velocities nearly collapse for $x_3^+ < 30$, while closer to the sphere at $x_1/D = 2$ and 4 large deviations are visible. Note that this is in stark contrast with the results for the sphere positioned at $h^+ = 306$ (Fig. 7).

3.3. *Effect of the sphere on the near wall Reynolds stress profiles*

Normalized, mean wall-normal profiles of one component of the Reynolds normal stress and the Reynolds shear stress are depicted in Figs. 9a and b, respectively, for $h^+ = 306$. The same is depicted in Fig. 10 for the sphere positioned closest to the wall.

In all cases, profiles strongly differ from those measured without the sphere. This is true in the wake of the sphere where stress magnitudes are strongly enhanced. Interestingly, there is quite a large effect on the stresses in the inner layer when the sphere is located at $h^+ = 306$ (Fig. 9). Of special interest is the position of the peak of the Reynolds shear stress that has moved away from the wall (from $x_3^+ \approx 30$ to 60) when the sphere is present. Comparing the stress profiles in Figs. 9 and 10 for different sphere

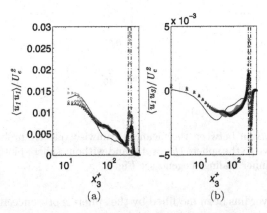

Fig. 9   Wall-normal profiles of normalized (a) Reynolds normal stress and (b) Reynolds shear stress, plotted in outer–inner scaling. $h^+ = 306$. For legend see Fig. 6.

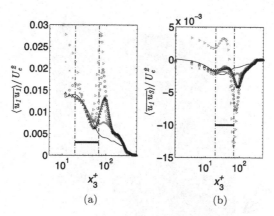

Fig. 10 Wall-normal profiles of (a) Reynolds normal stress and (b) Reynolds shear stress, plotted in outer–inner scaling. $h^+ = 43$. For legend see Fig. 6.

positions, it seems that they recover faster to those without sphere when the sphere is positioned closest to the wall.

3.4. *Instantaneous flow structures*

The present data were temporally resolved and instantaneous flow structures were visualized using the Q-criterion (Hunt *et al.*, 1988) that identifies vortices of an incompressible flow as connected fluid regions with a positive second invariant of the velocity gradient tensor:

$$Q \equiv \frac{1}{2}(\|\mathbf{\Omega}\|^2 - \|\mathbf{S}\|^2) > 0, \tag{1}$$

where $\mathbf{\Omega}$ and $\mathbf{S}$ are the anti-symmetric and symmetric parts of the velocity gradient tensor, respectively. Thus, the Q-criterion detects the regions where the vorticity magnitude prevails over the strain-rate magnitude.

Two examples of instantaneous vortex structures visualized by the Q-criterion are depicted in Figs. 11 and 12 for the sphere positioned at $h^+ = 43$ and 306, respectively. When the sphere is positioned furthest from the wall (Fig. 12), a clear distinction can be made between the large uplifted structure in the turbulent boundary layer and the small-scale structures in the sphere wake. In contrast, when the sphere is positioned closest to the wall (Fig. 11), wall-generated structures seem to be partially destroyed by the sphere. Note that in this case, the large uplifted wall generated structures extend to distances away from the wall beyond the extent of the sphere and as a result, vortical structures having spanwise rotation axes can be discerned just above the sphere's wake.

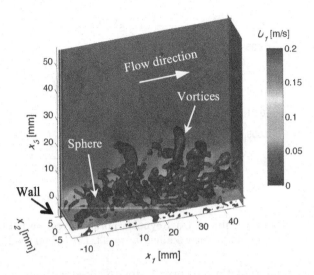

Fig. 11   Snapshot of instantaneous vortices visualized by the Q-criterion. Background contour plots denote the instantaneous streamwise velocity. $h^+ = 43$, $Re_D = 692$.

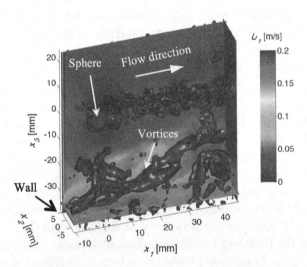

Fig. 12   Snapshot of instantaneous vortices visualized by the Q-criterion. Background contour plots denote the instantaneous streamwise velocity. $h^+ = 306$, $Re_D = 959$.

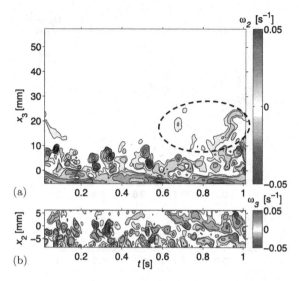

Fig. 13   Spatio-temporal plot of the out-of-plane vorticity components sampled at $x_1/D = 6$ and (a) $x_2/D = 0$, (b) $x_3/D = 0$. $h^+ = 43$.

Some idea of instantaneous interactions between the wall-generated turbulence and the sphere wake can be obtained from the spatio-temporal plots of out-of-plane components of the vorticity, ω_2 and ω_3, depicted in Figs. 13 and 14. These plots were obtained by sampling the instantaneous vorticity at a line positioned at $x_1/D = 6$. Figures 13a and 14a depict wall-normal planes centered at the sphere center, while Figs. 13b and 14b depict wall-parallel planes also centered at the sphere's center. As expected, the sphere wake at $h^+ = 306$ and $Re_D = 959$ (based on the mean velocity at the sphere's center) indicates a disordered pattern of co- and counter-rotating vortices as is expected at this Reynolds number (Sakamoto and Haniu, 1990).

The near-wall generated uplifted structures observed in the instantaneous snapshot depicted in Fig 12 can also be observed in Fig. 14 (dashed ellipse). This clearly shows a possible interaction mechanism between the sphere's wake and the large-scale coherent structures in the wall boundary layer.

When the sphere is positioned at $h^+ = 43$, it is clear that its wake is immersed in the boundary layer (Fig. 13a). Some pairs of counter-rotating vortices can be discerned, but in general the structure of the combined wake and boundary layer is quite disordered. As expected, near-wall coherent

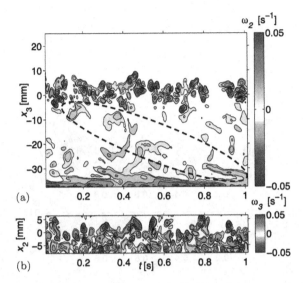

Fig. 14   Spatio-temporal plot of the out-of-plane vorticity components sampled at $x_1/D = 6$ and (a) $x_2/D = 0$, (b) $x_3/D = 0$. $h^+ = 306$.

structures are far less prominent, but those that penetrated into the outer layer remain visible (dashed ellipse in Fig. 13a). Clearly at $x_1/D = 6$, the boundary layer has not recovered as previously observed in the mean velocity and Reynolds stress profiles.

4. Concluding Remarks

Time-resolved tomo-PIV measurements were performed to investigate two-way coupling between a stationary held sphere ($D^+ = 51$) and a turbulent boundary layer (Re$_\theta$ = 752). Sphere Reynolds numbers were 692 and 959 when positioned at distances from the wall, $h^+ = 43$ and 306, respectively. At the present relatively low Re$_\theta$, the log layer extended up to $x_3^+ \approx 150$. Mean velocity and Reynolds stress profiles without the sphere were well resolved within the buffer, log, and outer layer. Introducing the sphere greatly changed both the mean velocity and Reynolds stress profiles across the whole boundary layer. When the sphere was positioned at $h^+ = 306$, recovery to the undisturbed boundary layer was slower than when the sphere was located at $h^+ = 43$. Furthermore, the peak of the Reynolds shear stress was shifted outward for $h^+ = 306$ for all measured streamwise positions. Instantaneous plots of vortical structures both in the sphere wake as well as those generated in the near-wall region show that large coherent, uplifted structures interact with the sphere's wake when positioned at $h^+ = 306$. When the sphere is positioned closest to the wall, the near-wall structures remain visible further away from it, while in the inner wall region, vortices shed from the sphere interact with the wall-bounded turbulence and the "undisturbed" boundary layer flow is not fully recovered at $x_1/D = 7$.

Acknowledgement

This research was supported by the Israel Science Foundation, grant no. 1596/14, and the JM Burgers center, Research School of Fluid Mechanics.

References

1. T. A. Johnson and V. C. Patel, "Flow past a sphere up to a Reynolds number of 300", *J. Fluid Mech.*, Vol. 378, pp. 19–70, (1999).
2. H. Sakamoto and H. Haniu, "A study on vortex shedding from spheres in a uniform flow", *J. Fluids Eng.*, Vol. 112, pp. 386–392, (1990).
3. C. Brücker, "Spatio-temporal reconstruction of vortex dynamics in axisymmetric wakes", *J. Fluids Struct.*, Vol. 15, pp. 543–554, (2001).
4. J.-S. Wu and G. M. Faeth, "Sphere wakes at moderate Reynolds numbers in a turbulent environment", *AIAA Journal*, Vol. 32, pp. 535–541, (1994).
5. P. Bagchi and S. Balachandar, "Effect of turbulence on the drag and lift of a particle", *Phy. Fluids*, Vol. 15, pp. 3496–3513, (2003).
6. L. Zeng, S. Balachandar, P. Fischer and F. Najjar, "Interactions of a stationary finite-sized particle with wall turbulence", *J. Fluid Mech.*, Vol. 594, pp. 271–305, (2008).
7. G. E. Elsinga and J. Westerweel, "Tomographic-PIV measurement of the flow around a zigzag boundary layer trip", *15th Int Symp on Applications of Laser Techniques to Fluid Mechanics*, Lisbon, Portugal, 05–08 July, (2010).
8. D. Adhikari and E. K. Longmire, "Visual hull method for tomographic PIV measurement of flow around moving objects", *Exp. Fluids*, Vol. 53, pp. 943–964, (2012).
9. G. E. Elsinga, R. J. Adrian, B. W. van Oudheusden and F. Scarano, "Three-dimensional vortex organization in a high-Reynolds-number supersonic turbulent boundary layer", *J. Fluid Mech.*, Vol. 644, pp. 35–60, (2010).
10. D. B. DeGraaff and J. K. Eaton, "Reynolds-number scaling of the flat-plate turbulent boundary layer", *J. Fluid Mech.*, Vol. 422, pp. 319–346, (2000).
11. L. P. Erm and P. N. Joubert, "Low-Reynolds-number turbulent boundary layers", *J. Fluid Mech.*, Vol. 230, pp. 1–44, (1991).
12. J. C. R. Hunt, A. A. Wray and P. Moin, "Eddies, streams and convergence zones in turbulent flows", *Center for Turbulence Research, Proceedings of the Summer Program*, pp. 193–208, (1988).

Breakup of Individual Colloidal Aggregates in Turbulent Flow Investigated by 3D Particle Tracking Velocimetry

Matthaus U. Babler[*,††], Alex Liberzon[†], Debashish Saha[‡],
Markus Holzner[§], Miroslav Soos[¶], Beat Lüthi[‖]
and Wolfgang Kinzelbach[**]

[*]*Department of Chemical Engineering,*
KTH Royal Institute of Technology,
SE-10044 Stockholm, Sweden
[†]*School of Mechanical Engineering,*
Tel Aviv University, Tel Aviv 69978, Israel
[‡]*Department of Applied Physics,*
Eindhoven University of Technology,
5600 MB, Eindhoven, The Netherlands
[§]*Environmental Fluid Mechanics,*
Institute of Environmental Engineering,
ETH Zurich, 8093 Zurich, Switzerland
[¶]*Department of Chemical Engineering,*
University of Chemistry and Technology, Technicka 3,
166 28 Prague 6 — Dejvice, Czech Republic
[‖]*Photrack AG, Ankerstr. 16a, 8004, Zurich, Switzerland*
[**]*Groundwater and Hydromechanics,*
Institute of Environmental Engineering,
ETH Zurich, 8093 Zurich, Switzerland
[††]*babler@kth.se*

Aggregates grown in mild shear flow are released, one at a time, into homogeneous isotropic turbulence where their breakup is recorded by three-dimensional particle tracking velocimetry (3D-PTV). The aggregates have an open structure with fractal dimension around 2.2, and their size varies from 0.9 to 3.1 mm which is large compared to the Kolmogorov length scale $\eta = 0.15$ mm. 3D-PTV allows for the simultaneous measurement of aggregate trajectories and the full velocity gradient tensor along their pathlines which enables us to access the Lagrangian stress history of individual breakup events. The analysis suggests that aggregates are mostly broken due to accumulation of drag stress over a time interval of order Kolmogorov time scale, $\mathcal{O}(\tau_\eta)$. This finding is explained by the fact that the aggregates are large, which gives their motion inertia and which increases the time for stress propagation inside the aggregate.

1. Introduction

Breakup of particle aggregates plays an important role in many industrial and natural applications, such as flocculation and coagulation in wastewater treatment [Biggs *et al.*, 2003; Fernandes *et al.*, 2015], microalgae production [Vandamme *et al.*, 2013], polymer manufacturing [Vaccaro *et al.*, 2006]; transport of sediments in river estuaries [Fugate and Friedrichs, 2003]; and drug delivery [Korin *et al.*, 2012]. Small aggregates in liquid media typically have small inertia and they follow the flow streamlines of the carrying fluid. This prevents the particles from undergoing impacts with solid objects such as vessel walls, turbine blades, or other particles. Moreover, collisions that do occur are strongly dampened due to lubrication effects which makes them unlikely to result in breakup [Synowiec *et al.*, 1993]. Therefore, breakup of small aggregate particles in liquid media is dominated by the hydrodynamic stress acting on the aggregate [Kusters, 1991; Babler *et al.*, 2015]. The particles that undergo breakup due to hydrodynamic stress are typically aggregates (or flocks) consisting of smaller entities called primary particles [Wengeler *et al.*, 2006]. The primary particles are held together by various forces whose exact origin depends on the specific applications. For instance, aggregates made of polymeric colloids, found in the manufacturing of materials, typically are held together by van der Waals forces that act between materials of the same kind [Ren *et al.*, 2015]. Another example is clay aggregates and aggregates of cellular material found in flocculation applications that involve polymeric binders that glue the particles together [Biggs *et al.*, 2003; Vandamme *et al.*, 2013].

Breakup of small aggregates is controlled by the interplay between the fluid flow in the surrounding of the aggregate that generates stress acting on the aggregate, and the mechanical response of the aggregate to the applied stress Kusters [1991]; Babler *et al.* [2008]; Vanni and Gastaldi [2011]; Vanni [2015]. The complexity of both these phenomena makes breakup a formidable problem. Simple mechanistic models based on dimensional arguments, which were successful in describing the opposite process of particle aggregation (i.e. the Smulochowski aggregation kernel Friedlander [2000]), proved to be inadequate to describe breakup. This becomes immediately clear when trying to derive the time scale of breakup: Considering an aggregate in a flow field, dimensional analysis provides us with a hydrodynamic time scale (e.g. the inverse of the characteristic velocity gradient on the length scale of the aggregate) that controls the stress acting the aggregate, and an aggregate time scale (e.g. the characteristic response time of the bond that holds the primary particles together) that controls the mechanical response of the aggregate. These two time scales are vastly different, and it is not clear how they combine to deliver the characteristic time for breakup Ó Conchúir and Zaccone [2013].

Additional complexity is added to the problem in the case where the flow suspending the aggregates is turbulent [Derksen, 2012; Babler *et al.*, 2012; De Bona *et al.*, 2014; Kobayashi *et al.*, 1999], which is the case in most applications (except, maybe, drug delivery [Korin *et al.*, 2012]). Turbulence affects the breakup process by creating strongly fluctuating hydrodynamic stresses acting on the aggregate everywhere in the bulk, even away from walls. Experimental and numerical approaches to study the breakup of aggregates in turbulence can be grouped roughly into two categories, i.e. approaches that consider individual aggregates and those considering ensembles (or populations) of aggregates. Studies that pursued the latter approach employed experiments with small aggregates that were measured by light scattering, while population balance modeling was used for the interpretation of the experiments [Soos *et al.*, 2006; Selomulya *et al.*, 2003; Flesch *et al.*, 1999]. Although in most of these studies, breakup occurred simultaneously with aggregation, the works provided valuable insights into the breakup process. It was established that breakup is size dependent, with larger aggregates breaking more easily than smaller ones. It was also found that breakup is a fast process with time scales that are significantly smaller than the characteristic times of aggregation, at least in the diluted conditions employed in the experiments.

However, the studies gave little insight into the actual mechanism of breakup, aspects which require to study individual aggregates and their interactions with the flow field. The fact that breakup is a fast process makes the study of individual aggregates a challenging endeavor. While the problem has received considerable interest from the numerical community, i.e. employing discrete element methods [Eggersdorfer *et al.*, 2010; Higashitani *et al.*, 2001; Becker *et al.*, 2009] and Stokesian dynamics [Harshe and Lattuada, 2012, 2016; Sanchez Fellaya and Vanni, 2012], there are only few studies [Blaser, 2000; Saha *et al.*, 2014; Glasgow and Hsu, 1982] that addressed the problem experimentally. The difficulty is not only the fast time scales of breakup but also the small length scales and the need for creating a flow field and aggregates that can be broken. While high-speed cameras and 3D-particle tracking provide a solution for the former issues, creating a controlled flow field and preparing aggregates that break inside the observation volume are challenges of their own kind.

In the present work, published in full in [Saha *et al.*, 2016], we addressed this challenge by tailoring both the aggregates and the flow field, i.e. we use a flow setup consisting of a "French washing machine" type of device that in its center generates turbulence that is close to homogeneous and isotropic. The aggregates are made of a fully destabilized polystyrene colloids and prepared in situ prior to the breakup experiment. By simultaneously monitoring the turbulent flow in the vicinity of the freely moving aggregate together with the aggregate itself, we are able to identify the hydrodynamic stress that prevails along the aggregate trajectory and at the point of breakup. The insight obtained from our experiments presents a significant step forward in the study of aggregate breakup in turbulence, and the employed techniques open the window for exploration of other flow configurations and related phenomena.

2. Experimental Setup and Methods

A schematic of the setup used to generate a turbulent flow is shown in Fig. 1 [Liberzon *et al.*, 2005]. The setup consists of a rectangular frame ($140 \times 120 \times 120$ mm^3) equipped with eight counter rotating disks (diameter 40 mm) mounted on opposite walls of the box. The frame is placed in a rectangular tank filled with deionized water at room temperature. The discharge of the rotating disks generates a flow that in the center of the frame is close to homogeneous and isotropic, validated by measuring the individual velocity components in x, y, and z-direction. The flow is seeded with a large number of tracer particles (polyamide, 100 μm in diameter) that are illuminated by a laser and tracked by a single high-speed camera (FASTCAM SA5, Photron USA, Inc. operated at 250 frames per second and a spatial resolution of 1024×1024 pixel^2) equipped with a four-view image splitter to obtain the view from the different angles Hoyer *et al.* [2005]. The observation volume where the particles are tracked is a rectangular box $30 \times 30 \times 40$ mm^3 placed in the center of the flow as shown in Fig. 1. In the present study, the disks revolved with 400 rpm which results in a turbulent flow that in observation volume assumes a Taylor scale Reynolds number of Re$_\lambda \approx 120$. The Kolmogorov length scale and time scale are $\eta = (\nu^3/\langle \varepsilon \rangle)^{1/4} = 0.15$ mm and $\tau_\eta = (\nu/\langle \varepsilon \rangle)^{1/2} = 0.02$ s (ν is the kinematic viscosity, $\langle \varepsilon \rangle$ is the

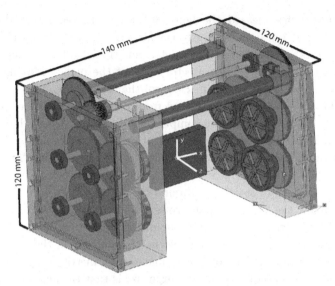

Fig. 1 Schematic of the flow setup used to generate homogeneous isotropic turbulence in which the aggregates are broken.

average energy dissipation rate in the observation volume). The integral length scale is $L = 25$ mm.

The aggregates are made of polystyrene particles with a diameter of 420 nm (Interfacial Dynamics Corp. Portland, OR, USA). The aggregates were prepared by mixing 0.3 mL of the original latex (solid volume fraction 8%) with 15 mL of deionized water. To this mixture, 15 mL of a 2.5 M NaCl is added to fully destabilize the latex. The mixture was sucked into a tube (PTFE, inner diameter 2 mm) that was connected to a syringe pump operated in oscillation mode with the piston moving back and forth at ca. 4 mm/min. The gentle oscillating flow generated by the syringe pump facilitates the formation of aggregates. The aggregates that formed in the PTFE tube were close to spherical with their size varying from 0.9 to 3.1 mm. The fractal dimension of the aggregate was measured by static light scattering and found to be $d_f \approx 2.2$, which is slightly higher than the typical value for RLCA aggregates. This means that the aggregates generated under the oscillatory flow had a slightly more compact structure than RCLA aggregates.

The experiments started from a stationary turbulent flow seeded with the tracer particles. To this flow, at each run, a single pre-formed aggregate was released using the PTFE tube used for the aggregate formation as feeding tube. Once released, the aggregates where tracked until and beyond breakup events. By simultaneously tracking the aggregate and the tracer particles in its surrounding, we can determine the local flow structure that prevails along the aggregate trajectory and at the point of breakup from which we aim at deducing the mechanism that causes breakup.

The local flow structure extracted from measuring the tracer particles in the vicinity of the aggregate comprise the local velocity gradient $A_{ij} = (\partial u_i/\partial x_j)$ and the slip velocity $q = u - v$, where u is the fluid velocity at the position of the aggregate and v is the aggregate velocity. The former was obtained by fitting a linear velocity profile to the tracer particles inside a spherical domain of size 3 mm around the aggregate.

To evaluate the hydrodynamic stress acting on the aggregate, we notice that the aggregate size is roughly one order larger than the Kolmogorov length scale. Hence, the aggregates do not fall entirely in the viscous subrange of the turbulent flow but are influenced by inertial range dynamics. The immediate consequence of this large size is that the aggregate motion is affected by its inertia. To quantify the aggregate inertia, we estimated the aggregate Stokes number, $St = \tau_p/\tau_d$, defined as the ratio of the characteristic time for acceleration of the aggregate, $\tau_p = \rho_a d_{\text{agg}}^2/(18\rho\nu)$ (ρ_a is the aggregate density taken as the density of polystyrene 1.05 g/cm³), and the time scale of the flow on the length scale of the aggregate, $\tau_d = (d_{\text{agg}}^2/\varepsilon)^{1/3}$ [Xu and Bodenschatz, 2008]. For aggregates with sizes $d_{\text{agg}}/\eta \in (6, 21)$ we found $St \in (0.6, 3.3)$, which implies pronounced deviations from fluid stream lines [Bec et al., 2006]. The latter have to be taken into account when estimating the hydrodynamic stress acting on the aggregate.

Here, we consider three stress contributions, namely the viscous shear stress caused by the velocity gradient on the length scale of the aggregate, the inertial stress (or normal stress) due to velocity differences on opposite side of the aggregate, and the drag stress due to the slip velocity of the aggregate. The viscous shear stress is estimated from the local energy dissipation rate as:

$$\sigma_{\text{shear}} = k_1\mu(\varepsilon/\nu)^{1/2}, \tag{1}$$

where k_1 is a prefactor that controls the transmission of stress to the aggregate structure, $\varepsilon = 2\nu s_{ij}s_{ij}$ is the local energy dissipation rate, s_{ij} is the rate of strain tensor, and μ is the dynamic viscosity. The normal stress is estimated from the velocity difference on two opposite sides of the aggregate, i.e. $\sigma_{\text{normal}} \sim \rho[\Delta u(d_{\text{agg}})]^2$, where $\Delta u(d_{\text{agg}})$ is the velocity difference over distance d_{agg}. For $d_{\text{agg}} > \eta$, the latter is $\Delta u \sim \rho(\varepsilon d_{\text{agg}})^{2/3}$, which results in:

$$\sigma_{\text{normal}} = k_2 \rho (\varepsilon d_{\text{agg}})^{2/3}, \tag{2}$$

where ρ is the fluid density. The drag stress is estimated from the drag force acting on the aggregate divided by the aggregate surface area (taken to be the surface of a sphere of diameter d_{agg}). Accounting for the aggregate inertia in the former this leads to:

$$\sigma_{\text{drag}} = k_3 \frac{3\mu|\boldsymbol{u} - \boldsymbol{v}|}{d_{\text{agg}}}(1 + 0.15\text{Re}_p^{0.687}), \tag{3}$$

where $Re_p = \rho|\boldsymbol{u} - \boldsymbol{v}|d_{\text{agg}}/\mu$ is the aggregate Reynolds number. The transmission factors k_1–k_3 in Eqs. (1)–(3) are complicated functions of the aggregate structure and the bonds that hold the primary particles together [Vanni and Gastaldi, 2011; Sanchez Fellaya and Vanni, 2012]. Relations for estimating these factors are not available such that, for simplicity, they are set to unity hereafter. We notice that the lack of numerical values for the transmission factors prevents us from comparing the relative magnitude of the different stress contributions.

3. Results and Discussion

In total, 80 aggregates were released and their trajectories were monitored until and beyond breakup. Figure 2 shows a series of snapshots from a random breakup experiment taken from one of the four views. The aggregate is shown by the large white spot while the smaller spots are the tracer particles used to visualize the flow (a video of the breakup experiments is available at "movie showing the breakup of an aggregate"[a]). Upon release,

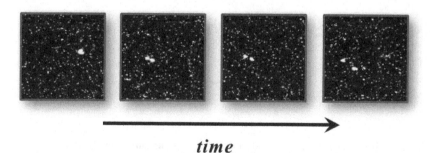

time

Fig. 2   Series of snapshots of the breakup of a random aggregate. The aggregate shown by the big blob is surrounded by tracer particles to visualize the flow.

---

[a]http://pubs.acs.org/doi/suppl/10.1021/acs.langmuir.5b03804

the aggregate gets engulfed by the turbulent flow which carries the aggregate away until it is broken up. It is noticed that the aggregates in our experiments indeed undergo breakup, i.e. the aggregate is broken into a few large fragments at a well-defined moment in time. This situation is different from the erosion case where small fragments (or primary particles) are continuously released from the aggregate surface.

By simultaneously measuring the aggregate and the tracer particles in its surrounding, we can determine the flow structure that prevails in the vicinity of the aggregate. From the latter, we can compute the hydrodynamic stresses along the aggregate trajectory using Eqs. (1)–(3). Figure 3 shows the time series of the hydrodynamic stress along the trajectory of three randomly selected aggregates. The aggregate is released at a time $t = 0$, and the time where it breaks is indicated by the vertical dashed

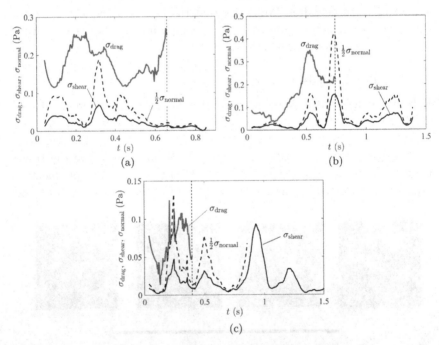

Fig. 3   Time series of the hydrodynamic stresses measured along the trajectory of individual aggregate. The aggregate is released at $t = 0$ while the moment of breakup is indicated by the vertical dashed line. The normal stress is scaled by a factor of $1/2$ for better visibility. For the cases shown in panel (a), (b) and (c) the aggregate size and aggregate Reynolds number are $d_{\mathrm{agg}} = 1.5$, 1.0, and 1.9 mm, respectively and $\mathrm{Re}_p = 45$, 21, and 31.

line. As can be seen, the hydrodynamic stresses acting on the aggregate are subject to strong fluctuations. Different scenarios for the occurrence of breakup were observed, i.e. in Fig. 3(a) breakup occurs at a local maximum of σ_{drag}, while in Fig. 3(b) breakup occurs at a global maximum of σ_{normal}. In the case shown in Fig. 3(c), breakup occurs right after the aggregate trajectory has passed a global maximum in the drag stress. This variation in the breakup pattern can be explained by differences in the specific aggregate structure, i.e. the arrangements of the primary particles within aggregate. Due to the way the aggregates were formed, the arrangement of the primary particles inside the aggregate is subject to strong variation which causes the different response to the applied stresses.

To determine which of the three stresses dominates the breakup, we explored the correlation between each stress contribution at the point of breakup and the aggregate size. The aggregate size crucially influences the aggregate strength, i.e. the strength decreases with increasing size, such that the correlation between size and stress provides a mean for identifying the dominant stress contribution. Evaluation of the three stress contributions (not shown) reveals that the drag stress shows the best correlation with the aggregate size. Figure 4(a) shows the drag stress at breakup plotted versus the inverse aggregate size in semi-logarithmic coordinates. Plotting the horizontal axis in this way corresponds to an aggregate strength that is increasing along the x-axis. Each point in Fig. 4(a) refers to a breakup event measured in the experiments. As can be seen, there is considerable

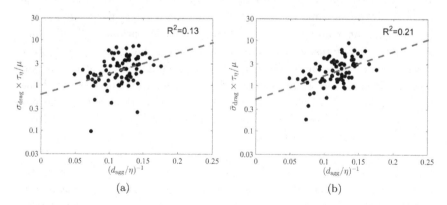

Fig. 4   Drag stress at the point of breakup (a) and accumulated drag stress (b) plotted vs. the inverse aggregate size. The accumulated drag stress in (b) is computed from Eq. (4) with $\Delta t = 3\tau_\eta$. The dashed lines indicate regression curves of the data, with the correlation coefficient (R^2) plotted in the panels.

scatter in the data, which reflects the variability of the fractal aggregates employed in this work. Nevertheless, the data shows a positive correlation between the inverse aggregate size and the drag stress at breakup, implying that a higher stress is needed to break a smaller (or stronger) aggregate.

To explore if breakup is controlled by the instantaneous stress acting on the aggregate or whether the stress history is relevant too [Blaser, 2000; Marchioli and Soldati, 2015], we evaluated the correlation between the inverse aggregate size and the average drag stress at breakup. The latter is defined as:

$$\bar{\sigma}_{\text{drag}} = \frac{1}{\Delta t} \int_{t_b - \Delta t}^{t_b} \sigma_{\text{drag}}(t) dt, \tag{4}$$

where Δt is a short time span over which the stress history influences the breakup and t_b is the breakup time. Fig. 4(b) shows the average drag stress at breakup plotted versus the inverse aggregate size. In the data shown, the time span was selected as $\Delta t \approx 3\tau_\eta$, which was found to give the best correlation. Comparing the instantaneous drag stress in Fig. 4(a) with the average drag stress shown in Fig. 4(b), it can be seen that the latter exhibits a significantly better correlation. From this, we conclude that it is the weak accumulation of the drag stress that controls the breakup of our aggregates.

4. Conclusions

We studied the breakup of aggregates in quasi-homogeneous and isotropic turbulence by means of 3D-PTV. The aggregates were made of a polystyrene latex and assumed a fractal dimension around 2.2. The large size of the aggregates with respect to the Kolmogorov length scale requires assessment of different stress contributions, namely shear stress, normal stress, and drag stress. Exploring different breakup criteria revealed that the accumulation of drag stress over a time lag on the order of the Kolmogorov time scale correlates best with the aggregate size. This profound result can be understood by taking into consideration the large size of the aggregates, with respect to both the primary particles and the smallest turbulent fluid motions. On one hand, the large size of the aggregates leads to a non-negligible inertia, which generates drag stress acting on the aggregate. On the other hand, the large size also increases the dissipation of stress inside the aggregate and the time it takes for the stress to be transmitted to the weakest link inside the aggregate, such that, in effect, stress must be applied over a longer time to break the aggregate.

Acknowledgements

This work was financially supported by the Swiss National Science Foundation (Grant No. 119815). M.U.B. was financially supported by the Swedish Research Council VR (Grant No. 2012-6216). M.H. acknowledges support from SNSF Grant No. 144645. M.S. was partially supported by the Specific University Research grant of UCT (Grant No. 20/2015). EU-COST action MP1305 is kindly acknowledged.

References

[1] C. Biggs, P. Lant and M. Hounslow, "Modelling the effect of shear history on activated sludge flocculation", *Water Sci. Tech.*, Vol. 47, pp. 251–257, (2003).

[2] A. Fernandes, Y. Lawryshyn, J. Gibson and R. Farnood, "Experimental and numerical investigation of the breakage of wastewater flocs in orifice flow", *Wat. Qual. Res. J. Can.*, Vol. 50, pp. 47–57, (2015).

[3] D. Vandamme, I. Foubert and K. Muylaert, "Flocculation as a low-cost method for harvesting microalgae for bulk biomass production", *Trends Biotechnol.*, Vol. 31, pp. 233–239, (2013).

[4] A. Vaccaro, J. Sefcik, H. Wu, M. Morbidelli, J. Bobet and C. Fringant, "Aggregation of concentrated polymer latex in stirred vessels", *AIChE J.*, Vol. 52, pp. 2742–2756, (2006).

[5] D. C. Fugate and C. T. Friedrichs, "Controls on suspended aggregate size in partially mixed estuaries", *Estuarine Coastal Shelf Sci.*, Vol. 58, pp. 389–404, (2003).

[6] N. Korin, M. Kanapathipillai, B. D. Matthews, A. Crescente, M. Brill, T. Mammoto, K. Ghosh, S. Jurek, S. A. Bencherif, D. Bhatta, A. U. Coskun, C. L. Feldman, D. D. Wagner and D. E. Ingber, "Shear-Activated Nanotherapeutics for Drug Targeting to Obstructed Blood Vessels", *Science*, Vol. 337, pp. 737–742, (2012).

[7] P. Synowiec, A. G. Jones and P. A. Shamlou, "Crystal break-up in dilute turbulently agitated suspensions", *Chem. Eng. Sci.*, Vol. 48, pp. 3485–3495, (1993).

[8] K. A. Kusters, "The influence of turbulence on aggregation of small particles in agitated vessels", PhD Thesis, Technische Universiteit Eindhoven, (1991).

[9] M. U. Babler, L. Biferale, L. Brandt, U. Feudel, K. Guseva, A. S. Lanotte, C. Marchioli, F. Picano, G. Sardina, A. Soldati and F. Toschi, "Numerical simulations of aggregate breakup in bounded and unbounded turbulent flows", *J. Fluid Mech.* Vol. 766, pp. 104–128, (2015).

[10] R. Wengeler, A. Teleki, M. Vetter, S. E. Pratsinis and H. Nirschl, "High-pressure liquid dispersion and fragmentation of flame-made silica agglomerates", *Langmuir* Vol. 22, pp. 4928–4935, (2006).

[11] Z. Ren, Y. M. Harshe and M. Lattuada, "Influence of the potential well on the breakage rate of colloidal aggregates in simple shear and uniaxial extensional flows", *Langmuir*, Vol. 31, pp. 5712–5721, (2015).

[12] M. U. Babler, M. Morbidelli and J. Baldyga, "Modelling the breakup of solid aggregates in turbulent flows", *J. Fluid Mech.*, Vol. 612, pp. 261–289, (2008).

[13] M. Vanni and A. Gastaldi, "Hydrodynamic forces and critical stresses in low-density aggregates under shear flow", *Langmuir*, Vol. 27, pp. 12822–12833, (2011).

[14] M. Vanni, "Accurate modelling of flow induced stresses in rigid colloidal aggregates", *Comput. Phys. Commun.*, Vol. 192, pp. 70–90, (2015).

[15] S. K. Friedlander, *Smoke, Dust, and Haze: Fundamentals of Aerosal Behavior*, Oxford University Press, 2000.

[16] B. Ó Conchúir and A. Zaccone, "Mechanism of flow-induced biomolecular and colloidal aggregate breakup", *Phys. Rev. E*, Vol. 87, pp. 032310, (2013).

[17] J. J. Derksen, "Direct numerical simulations of aggregation of monosized spherical particles in homogeneous isotropic turbulence", *AIChE J.*, Vol. 58, pp. 2589–2600, (2012).

[18] M. U. Babler, L. Biferale and A. S. Lanotte, "Breakup of small aggregates driven by turbulent hydrodynamical stress", *Phys. Rev. E*, Vol. 85, p. 025301, (2012).

[19] J. De Bona, A. S. Lanotte and M. Vanni, "Internal stresses and breakup of rigid isostatic aggregates in homogeneous and isotropic turbulence", *J. Fluid Mech.* Vol. 755, pp. 365–396, (2014).

[20] M. Kobayashi, Y. Adachi and O. Setsuo, "Breakup of fractal flocs in a turbulent flow", *Langmuir*, Vol. 15, pp. 4351–4356, (1999).

[21] M. Soos, J. Sefcik and M. Morbidelli, "Investigation of aggregation, breakage and restructuring kinetics of colloidal dispersions in turbulent flows by population balance modeling and static light scattering", *Chem. Eng. Sci.*, Vol. 61, pp. 2349–2363, (2006).

[22] C. Selomulya, G. Bushell, R. Amal and T. D. Waite, "Understanding the role of restructuring in flocculation: The application of a population balance model", *Chem. Eng. Sci.*, Vol. 58, pp. 327–338, (2003).

[23] J. C. Flesch, P. T. Spicer and S. E. Pratsinis, "Laminar and turbulent shear-induced flocculation of fractal aggregates", *AIChE J.*, Vol. 45, pp. 1114–1124, (1999).

[24] M. L. Eggersdorfer, D. Kadau, H. J. Herrmann and S. E. Pratsinis, "Fragmentation and restructuring of soft-agglomerates under shear", *J. Colloid Interface Sci.*, Vol. 342, pp. 261–268, (2010).

[25] K. Higashitani, K. Iimura and H. Sanda, "Simulation of deformation and breakup of large aggregates in flows of viscous fluids", *Chem. Eng. Sci.*, Vol. 56, pp. 2927–2938, (2001).

[26] V. Becker, E. Schlauch, M. Behr and H. Briesen, "Restructuring of colloidal aggregates in shear flows and limitations of the free-draining approximation", *J. Colloid Interface Sci.*, Vol. 339, pp. 362–372, (2009).

[27] Y. M. Harshe and M. Lattuada, "Breakage rate of colloidal aggregates in shear flow through Stokesian dynamics", *Langmuir*, Vol. 28, pp. 283–292, (2012).

[28] Y. M. Harshe and M. Lattuada, "Universal breakup of colloidal clusters in simple shear flow", *J. Phys. Chem. B*, Vol. 120, pp. 7244–7252, (2016).

[29] L. Sanchez Fellaya and M. Vanni, "The effect of flow configuration on hydrodynamic stresses and dispersion of low density rigid aggregates", *J. Colloid Interface Sci.* Vol. 388, pp. 47–55, (2012).

[30] S. Blaser, "Break-up of flocs in contraction and swirling flows", *Colloid Surface A*, Vol. 166, pp. 215–223, (2000).

[31] D. Saha, M. Soos, B. Lüthi, M. Holzner, A. Liberzon, M. U. Babler and W. Kinzelbach, "Experimental characterization of breakage rate of colloidal aggregates in axisymmetric extensional flow", *Langmuir*, Vol. 30, pp. 14385–14395, (2014).

[32] L. A. Glasgow and J. P. Hsu, "An experimental study of floc strength", *AIChE J.*, Vol. 28, pp. 779–785, (1982).

[33] D. Saha, M. U. Babler, M. Holzner, M. Soos, B. Lü thi, A. Liberzon and W. Kinzelbach, "Breakup of finite-size colloidal aggregates in turbulent flow investigated by three-dimensional (3D) particle tracking velocimetry", *Langmuir*, Vol. 32, pp. 55–65, (2016).

[34] A. Liberzon, M. Guala, B. Lüthi, W. Kinzelbach and A. Tsinober, "Turbulence in dilute polymer solutions", *Phys. Fluids*, Vol. 17, p. 031707, (2005).

[35] K. Hoyer, M. Holzner, B. Lüthi, M. Guala, A. Liberzon and W. Kinzelbach, "3D scanning particle tracking velocimetry", *Exp. Fluids*, Vol. 39, pp. 923–934, (2005).

[36] H. Xu and E. Bodenschatz, "Motion of inertial particles with size larger than Kolmogorov scale in turbulent flows", *Phys. D*, Vol. 237, pp. 2095–2100, (2008).

[37] J. Bec, L. Biferale, G. Boffetta, A. Celani, M. Cencini, A. S. Lanotte, S. Musacchio and F. Toschi, "Acceleration statistics of heavy particles in turbulence", *J. Fluid Mech.*, Vol. 550, pp. 349–358, (2006).

[38] C. Marchioli and A. Soldati, "Turbulent breakage of ductile aggregates", *Phys. Rev. E*, Vol. 91, p. 053003, (2015).

Structural Stability and Transient Changes of the Liquid Level in Stratified Flow

Yehuda Taitel[*] and Dvora Barnea[†]

*School of Mechanical Engineering,
Tel Aviv University, Tel-Aviv 69978, Israel*
*taitel@eng.tau.ac.il
†dbarnea@eng.tau.ac.il

Substantial changes of the liquid holdup along the pipe may take place due to small changes in gas or liquid flow rates. These changes can be attributed to the existence of multiple steady state solutions in stratified flow. It is aimed here to investigate the structural stability of stratified flow by applying the transient two-fluid model using the method of characteristics and to compare it to a simplified model used in the past for the structural stability analysis. It is shown here that the two approaches for structural stability analysis are consistent and in fact complement each other. The transient simulations performed with the two-fluid model have the advantage of following the transient changes in the liquid level along the pipe.

1. Introduction

A comprehensive review on various approaches for stability analyses of separated flows (stratified or annular) is presented by Taitel and Barnea (2016). In that review, the authors focused on two kinds of instabilities: (1) interfacial instability, which indicates whether the interface is unstable resulting in a wavy interface and (2) structural instability which indicates whether the structure of separated flow is stable with respect to the average film thickness.

One may get a stable separated flow structure with an unstable wavy interface like in annular flow, or a stable smooth interface but an unstable structure for certain cases of stratified flow. Both approaches (the interfacial and structural stabilities) were analyzed linearly and nonlinearly (Barnea,

1991; Landman, 1991; Barnea and Taitel, 1992, 1993, 1994a, 1994b). The results were applied either to detect unrealistic steady state solutions when multiple solutions take place or to identify possible instabilities that may lead to transition from stratified flow or from annular flow.

We will first briefly review the interfacial and the structural stability analyses that were presented in the past. In this paper, we will show that the results of the structural stability analyses (linear and nonlinear) that were based on some simplified assumptions are confirmed by the more rigorous two-fluid model. Using this approach, one can follow the development of the stratified liquid layer profile with time leading to steady flow.

2. Analysis

The analysis of the stability of stratified flow is based here on the two-fluid model where transient continuity and momentum equations are applied to gas and liquid together with the appropriate constitutive relations (Andritsos et al., 1989; Barnea, 1991; Brauner and Moalem-Maron, 1991; Barnea and Taitel, 1993). Assuming incompressible isothermal flow with no mass transfer yields the following equations:

Continuity of the liquid

$$\frac{\partial h_L}{\partial t} + U_L \frac{\partial h_L}{\partial x} + H_L \frac{\partial U_L}{\partial x} = 0. \tag{1}$$

Continuity of the gas

$$\frac{\partial h_L}{\partial t} + U_G \frac{\partial h_L}{\partial x} - H_G \frac{\partial U_G}{\partial x} = 0. \tag{2}$$

The combined momentum equation

$$(\rho_L - \rho_G)g\cos\beta \frac{\partial h_L}{\partial x} + \rho_L \frac{\partial U_L}{\partial t} - \rho_G \frac{\partial U_G}{\partial t}$$

$$+\rho_L U_L \frac{\partial U_L}{\partial x} - \rho_G U_G \frac{\partial U_G}{\partial x} - \sigma \frac{\partial^3 h_L}{\partial x^3} = F, \tag{3}$$

where

$$F = -\frac{\tau_L S_L}{A_L} + \frac{\tau_G S_G}{A_G} + \tau_i S_i \left(\frac{1}{A_L} + \frac{1}{A_G}\right) - (\rho_L - \rho_G)g\sin\beta. \tag{4}$$

The shear stresses in Eq. (4) are evaluated by:

$$\tau_L = f_L \frac{\rho_L U_L^2}{2} \quad \tau_G = f_G \frac{\rho_G U_G^2}{2} \quad \tau_i = f_i \frac{\rho_G (U_G - U_L) |U_G - U_L|}{2}. \tag{5}$$

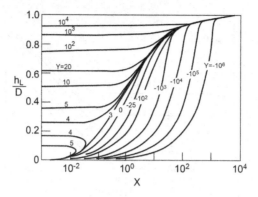

Fig. 1 Equilibrium liquid level for stratified flow, turbulent liquid, turbulent gas.

A is the cross sectional area, U is the axial average velocity, τ is the wall shear stress, τ_i is the interfacial shear stress, S and S_i are the perimeters over which τ and τ_i act, h_L is the liquid level, ρ is density, σ is surface tension, and β is the inclination angle from the horizontal. The subscripts L and G refer to the liquid and gas respectively, $H_L = A'_L/A_L$ where $A'_L = dA_L/dh_L$. The interfacial friction factor, f_i, was assumed constant having a value of 0.014 (Cohen and Hanratty, 1968).

For the case of steady uniform flow, Eq. (3) reduces to $F = 0$, which yields the steady state solutions.

Taitel and Dukler (1976) showed that all terms in F can be expressed as a function of the liquid level h_L/D, the Lockhart–Martinelli parameter, X, and the inclination parameter Y.

$$X = \left[\frac{(dP/dx)_{LS}}{(dP/dx)_{GS}} \right]^{1/2} \qquad Y = \frac{(\rho_L - \rho_G)g\sin\beta}{|(dP/dx)_{GS}|}$$

U_{LS} and U_{GS} are the liquid and gas superficial velocities. $(dP/dx)_{LS}$ and $(dP/dx)_{GS}$ are the pressure drops of the liquid and gas, respectively, flowing alone in the pipe.

Figure 1 presents the steady state solutions. One can observe that there is a range where multiple solutions exist.

In Fig. 2, the solutions for h_L/D are presented using dimensional coordinates for upward inclination angles of 0.25° and 1°. In the upper figures, the steady state liquid level, h_L/D, is plotted vs. U_{LS} for various values of U_{GS}, and in the bottom figure h_L/D is plotted vs. U_{GS} for several values of U_{LS}. It can be seen that multiple solutions are obtained for a wide range of liquid flow rates and a narrow range of gas flow rates.

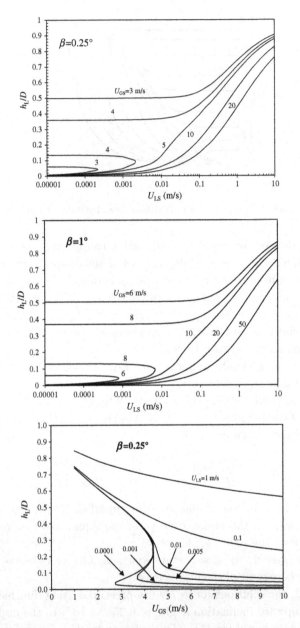

Fig. 2   Steady state solutions, air–water, 0.1 MPa, 25°C, 5 cm dia. pipe.

3. Interfacial Stability

A linear Kelvin–Helmhotz stability analysis yields the following conditions for the interfacial stability (Barnea and Taitel, 1992).

The viscous Kelvin–Helmholz (VKH) analysis which includes the effect of viscosity yields

$$(C_V - C_{IV})^2 + \frac{\rho_L \rho_G}{\left(\frac{\rho_L}{R_L} + \frac{\rho_G}{R_G}\right)^2 R_L R_G} (U_G - U_L)^2$$

$$-\frac{(\rho_L - \rho_G) g \cos\beta}{\frac{\rho_L}{R_L} + \frac{\rho_G}{R_G}} \frac{A}{A'_L} - \frac{A}{A'_L} \frac{\sigma}{\left(\frac{\rho_L}{R_L} + \frac{\rho_G}{R_G}\right)} \left(\frac{2\pi}{\lambda}\right)^2 \leq 0. \qquad (6)$$

The last three terms on the L.H.S. of (6) comprise the stability criterion for the inviscid Kelvin–Helmholtz (IKH) analysis where the viscous effects are neglected. The first term is the additional effect of the shear stresses.

C_V in is the critical wave velocity on the inception of instability

$$C_V = \frac{\left(\frac{\partial F}{\partial R_L}\right)_{U_{GS}, U_{LS}}}{\left[\left(\frac{\partial F}{\partial U_{GS}}\right)_{U_{LS}, R_L} - \left(\frac{\partial F}{\partial U_{LS}}\right)_{U_{GS}, R_L}\right]} \qquad (7)$$

and C_{IV} is the critical wave velocity for the IKH analysis

$$C_{IV} = \frac{\rho_L U_L R_G + \rho_G U_G R_L}{\rho_L R_G + \rho_G R_L}, \qquad (8)$$

Where R_L and R_G are the liquid and gas holdups and λ is the wavelength.

The dashed (blue) line in Fig. 3 is the IKH neutral stability curve for the case of long waves where $\sigma = 0$. The solutions in the region outside this curve are associated with an unbounded growth of any disturbance, resulting in transition from stratified flow (Barnea, 1991; Barnea and Taitel, 1993). This region is also associated with ill posedness of the formulation (Brauner and Moalem-Marom, 1991). The thin (red) lines within the region bounded by the dashed (blue) line are solutions that are unstable to the VKH analysis, resulting in a wavy interface or transition from stratified flow (when the waves are large and touch the upper wall). The thick (black) lines are stable solutions resulting in smooth stratified flow. Note that when we have multiple solutions, the first (thinnest) solution is stable to VKH, the third solution is unstable, and the second solution may be stable or unstable. For a pipe inclination of 0.25°, the second solution is interfacially stable, while for an inclination angle of 1.0° the second solution is unstable.

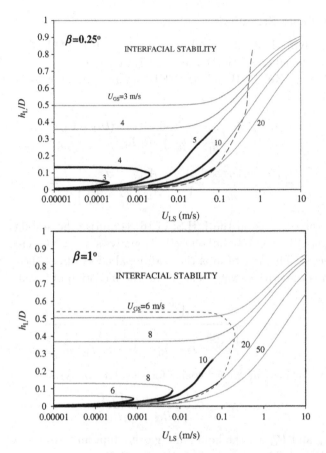

Fig. 3   Interfacial linear stability, air–water, 0.1 MPa, 25°C, 5 cm dia. pipe.
Dashed (blue) curves — neutral stability, IKH; thick (black) curves — stable
to VKH; thin (red) curves — unstable to VKH; gray curves — unstable to
IKH & VKH.

Barnea and Taitel (1994a) also investigated the nonlinear interfacial
stability using numerical simulations to examine the system response to
a finite wavy disturbance on the interface, following the propagation and
evolution of the wavy disturbance. The results of the nonlinear simulation
are consistent with the results of the linear analysis.

4.   Structural Stability — Simplified

The interfacial stability analysis is not sufficient to determine the stability of separated flow. For example, steady annular flow is unstable with

respect to its interface, resulting in a wavy interface but still annular flow is a stable structure with respect to its average film thickness. In addition, for the case of cocurrent and countercurrent annular flow, multiple steady state solutions are obtained where some of them are not stable to their structure. Similarly for stratified flow, the interface may be stable, resulting in a smooth interface but some of the solutions may be unstable with respect to their structure and thus will not exist. Barnea and Taitel (1992) suggested examining the stability of the structure by using a simplified analysis assuming that the liquid level is uniform along the pipe and it varies only with time. In this case, the momentum equation for the liquid is

$$\frac{dU_L}{dt} = -\frac{U_{LS}}{R_L \ell} \left(U_L - \frac{U_{LS}}{R_L} \right) + \frac{F}{\rho_L} \tag{9}$$

and the continuity equations is

$$\frac{dR_L}{dt} = \frac{1}{\ell} \left(U_{LS} - U_L R_L \right), \tag{10}$$

where ℓ is the pipe length.

Linear stability analysis yields the following criterion for the structural stability (Barnea and Taitel, 1992)

$$
\begin{aligned}
\left(\frac{\partial F}{\partial R_L} \right)_{U_{LS}, U_{GS}} &> 0 \quad \text{for cocurrent flow,} \\
\left(\frac{\partial F}{\partial R_L} \right)_{U_{LS}, U_{GS}} &< 0 \quad \text{for counter-current flow.}
\end{aligned}
\tag{11}
$$

For the case of cocurrent stratified flow, when we have multiple solutions, the first and the third solutions are structurally stable according to the linear analysis (Eq. (11)), while the second steady state solution is linearly unstable. The response of the system (Eqs. (9) and (10)) to finite changes of the average film thickness indicates which steady state solution is nonlinearly stable. Figure 4 shows the results for the nonlinear analysis for 0.25° and 1° inclination angles, respectively. The second solution which is linearly unstable according to criterion (11) is also unstable to finite disturbances and the trajectories run away from the second solution to solutions 1 and 3 which are linearly stable. However, the trajectories towards the third solution pass through negative liquid flow velocities, which can be interpreted as destruction of the cocurrent stratified flow resulting in transition from stratified flow. The solid thick (black) lines in Fig. 5 are structurally stable solutions for infinitesimal disturbance (linear analysis). The solid thin (red) curves present solutions that are linearly unstable to the structure. The dotted (red) curves are unstable solutions to finite disturbances. Note

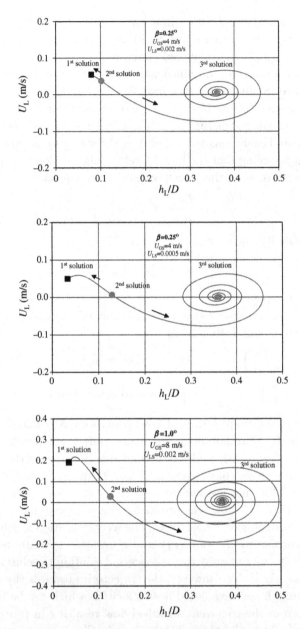

Fig. 4   Nonlinear structural stability — simplified model, $\ell = 1\,\text{m}$, $D = 5\,\text{cm}$. ■: first solution, ●: second solution, ▲: third solution.

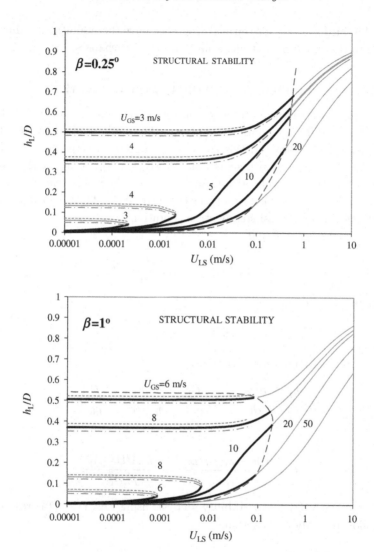

Fig. 5 Structural stability. Thick full (black) curves — structurally stable solutions — linear analysis; thin full (red) curves — structurally unstable solutions — linear analysis; dotted curves — nonlinear structurally unstable solutions, simplified model simulations; dashed/dot curves — nonlinear structurally unstable solutions, simulation using TFM; dashed (blue) curves — neutral stability IKH.

that the third solution for the case where we have three solutions is linearly stable to the structure but unstable to finite disturbances.

5. Structural Stability — Method of Characteristics

The nonlinear structural stability analysis is studied here using the two-fluid model (TFM) equations applying the method of characteristics. The method of characteristics cannot be applied directly on the TFM, Eqs. (1)–(4) since the system is not strictly hyperbolic as some characteristics have infinite velocities (also for $\sigma = 0$). Therefore, the method of characteristics was applied to a transient form of the TFM, which assumes that the gas is in a quasi steady state condition.

Continuity of the liquid

$$\frac{\partial h_L}{\partial t} + U_L \frac{\partial h_L}{\partial x} + H_L \frac{\partial U_L}{\partial x} = 0. \tag{12}$$

Continuity of the gas

$$A_G U_G = A U_{GS}. \tag{13}$$

The combined momentum equation is

$$G \frac{\partial h_L}{\partial x} + \frac{\partial U_L}{\partial t} + U_L \frac{\partial U_L}{\partial x} - \frac{\sigma}{\rho_L} \frac{\partial^3 h_L}{\partial x^3} - \frac{F}{\rho_L} = 0, \tag{14}$$

where

$$G = \frac{(\rho_L - \rho_G) g \cos \beta}{\rho_L} - \frac{\rho_G A^2 U_{GS}^2 A_L'}{\rho_L A_G^3}. \tag{15}$$

Note that the VKH linear stability analysis using Eqs. (12)–(14) results in the following stability criterion:

$$\left[\frac{\left(-\frac{\partial F}{\partial R_L} \right)_{U_{LS}}}{\left(\frac{\partial F}{\partial U_{LS}} \right)_{R_L}} - U_L \right]^2 + \frac{\rho_G U_{GS}^2 R_L}{\rho_L R_G^3}$$

$$- \frac{(\rho_L - \rho_G) g \cos \beta \frac{A}{A_L'} R_L}{\rho_L} - \frac{A}{A_L'} \frac{\sigma}{\rho_L} R_L \left(\frac{2\pi}{\lambda} \right)^2 \leq 0 \tag{16}$$

which yields almost the same results as those of Eq. (6) (presented in Fig. 3) (Barnea and Taitel, 1994a).

Equations (12)–(14) for $\sigma = 0$, can be converted into two ordinary differential equations along two characteristic directions as follows:

$$\frac{dh_L}{dt} \pm \sqrt{\frac{H_L}{G}} \frac{dU_L}{dt} \pm \sqrt{\frac{H_L}{G}} \left(-\frac{F}{\rho_L} \right) = 0 \quad \text{along} \quad \frac{dx}{dt} = U_L \pm \sqrt{GH_L}. \quad (17)$$

Barnea and Taitel (1994a) used the method of characteristics to check the nonlinear interfacial instability by introducing a wavy disturbance on the steady interface, following the evolution (growth or decay) of the wave.

In the present analysis, we use the method of characteristics to check the structural stability by applying a disturbance on the whole- film level (contrary to a wavy disturbance) and follow the evolution of the liquid level with time. When the steady state solution yields supercritical flow, it is more convenient to use finite differences method (implicit upwind discretization).

Figures 6 and 7 show some examples of transient simulations for the case where the inclination angles are $\beta = 0.25°$ and $\beta = 1.0°$, respectively.

Figure 6 considers a case having three steady state solutions. The first and the third solutions are linearly stable to the structure and the second is linearly unstable (Eq. (11)). In Fig. 6(a), an initial liquid level somewhat below the second solution is applied, and the evolution of the level profile is shown for different times. It is clearly seen that the liquid level approaches the stable first solution and runs away from the second unstable solution. Figure 6(b) shows that when the disturbance in the level is below the first stable solution, the trajectory approaches the stable steady first solution. Note that at steady state, the liquid level is uniform along the pipe. Figure 6(c) shows the case where the initial level is somewhat below the third solution, which is linearly stable to the structure. It seems that the solution tries to approach the linearly stable third solution, but it develops severe oscillations, indicating that the third solution is unstable to finite disturbances.

Figure 7 presents a similar case when the angle of inclination is $1.0°$. Unlike the previous case for $\beta = 0.25°$ where the flow conditions are subcritical, for $\beta = 1.0°$ the first steady state solution is supercritical. In this case, it is more convenient to use a finite difference method (implicit upwind discretization). Figures 7(a) and 7(b) show that after a finite disturbance below and above the first solution, which is linearly stable to the structure, the trajectories approach in time the stable first solution. Figure 7(c) shows the profile evolution when the initial level is somewhat below the

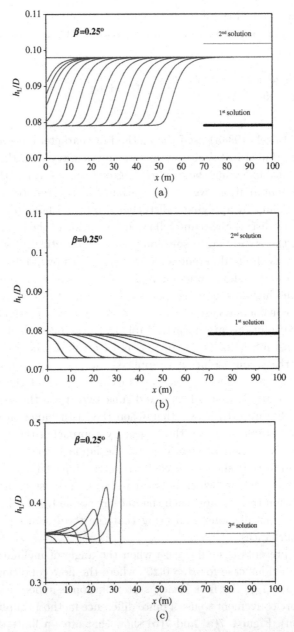

Fig. 6   Transient changes of the liquid level owing to a disturbance in the liquid
level around the three steady state solutions. $\beta = 0.25°$, $U_{LS} = 0.002\,\mathrm{m/s}$, $U_{GS} = 4.0\,\mathrm{m/s}$.

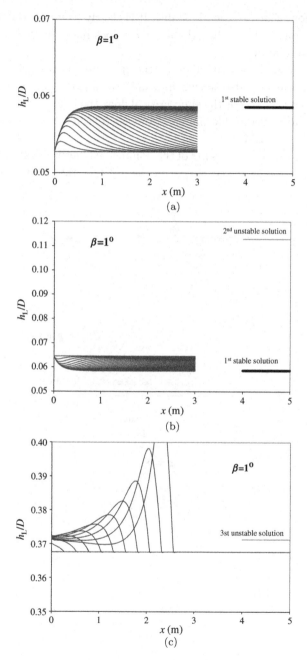

Fig. 7 Transient changes of the liquid level owing to a disturbance in the liquid level around the three steady state solutions. $\beta = 1°$, $U_{LS} = 0.005\,\text{m/s}$, $U_{GS} = 8.0\,\text{m/s}$.

third solution, which is linearly stable. It is clearly seen that the third solution is unstable to finite disturbances as it does not converge to the third steady state solution.

For the case of subcritical flow, the flow rate is prescribed and the inlet level is calculated resulting in a uniform liquid level at steady state (Figs. 6(a) and 6(b)). For the supercritical flow, the inlet level is prescribed, thus the steady level profile changes from the inlet level to the steady state level (Figs. 7(a) and 7(b)).

Figure 8 presents the results of the nonlinear structural stability analysis obtained by applying disturbances in the gas flow rate, using the method

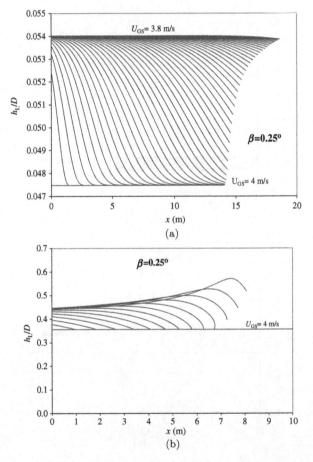

Fig. 8   Transient changes of the liquid level owing to a decrease in gas flow rate. $\beta = 0.25°$, $U_{LS} = 0.001$ m/s: (a) $U_{GS} = 4 \to 3.8$ m/s, (b) $U_{GS} = 4 \to 3.5$ m/s.

of characteristics. Figure 8(a) presents the evolution of the liquid level due to a decrease of the gas superficial velocity from $U_{GS} = 4\,\mathrm{m/s}$ to $3.8\,\mathrm{m/s}$ for $U_{LS} = 0.001$ along the first solution. Since the initial and the final states are stable, the level profile changes smoothly. A decrease of U_{GS} from $4\,\mathrm{m/s}$ which starts at the third solution to $3.5\,\mathrm{m/s}$ where only a single solution exists (see Fig. 4(c)) indicates that the solution diverges, namely, the solution is structurally unstable although it is a single solution below the critical gas velocity, where only single solutions exist.

Figure 5 presents a summary of the results for the structural stability analyses, linear and nonlinear, using the simplified analysis (Eqs. (9) and (10)) and the characteristic analysis (Eq. (17)). The thick full (black) curves are structurally stable solutions as obtained by the linear stability analysis (Eq. (11)) and the thin full (red) curves are structurally linearly unstable solutions. The first and the third solutions are linearly stable, while the second solution is linearly unstable.

The nonlinear structural stability analysis based on the simplified model (Eqs. 9, 10) confirms that the 2nd solution is unstable and the transient trajectory of the liquid level runs away towards the first stable solution or to the third solution. However, the convergence to the third solution passes through negative velocities, which are interpreted as structurally unstable (Fig. 4). These results are plotted in Fig. 5 as dotted (red) curves.

The dashed/dot (red) curves are nonlinearly unstable solutions obtained using the TFM numerical simulations. Note that the results obtained by this method also show that the third solution is unstable to finite disturbances (in the liquid level or in the flow rates) and the liquid level diverges with time. This method becomes ill posed as the liquid flow rate increases. It takes place very closely to the boundary where the flow is unstable to the IKH (the gray lines).

As seen, the results based on the simplified approach (assuming uniform liquid level) are very similar to the one based on the TFM using characteristic analysis, yet the TFM approach provides additional information on the transient variation of the liquid level along the pipe.

6. Summary

In upward gas liquid stratified flow, one may get multiple steady state solutions for the liquid level. It is necessary to determine which solution will actually occur and to check whether it is possible to get two or more solutions for the same operational conditions.

It is well known that interfacial stability does not uniquely determine the existence of stratified flow. The interface may be unstable and wavy, and stratified flow will exists. Conversely, the interface may be stable and smooth, and still stratified flow will not take place. Thus, interfacial stability analysis is not sufficient to determine the existence of stratified flow requiring consideration of the structural stability.

The stability of the stratified flow structure is analyzed using (a) a simplified analysis assuming the liquid level is uniform along the pipe and it is a function of time only and (b) a TFM that allows obtaining the liquid level profiles with time, using the method of characteristic.

It is shown that the nonlinear analysis results obtained by the simplified approach and the method of characteristics are consistent with the results of the linear analysis related to the first and the second solutions. The third solution, which is linearly stable, (Eq. (11)), is unstable to the nonlinear analyses obtained by the two approaches: (a) the simplified method (Eqs. (9) and (10)) where the instability is reflected by the negative velocity (the dotted red lines) and (b) the TFM approach using the method of characteristics where the instability is reflected by the divergence of the liquid level up to the point where the system is ill posed (the dashed-dot red lines). The results of the TFM approach confirm the validity of the simplified approach.

Acknowledgment

Support from the Israeli Science Foundation, grant # 281/14 is gratefully acknowledged

References

N. Andritsos, L. Williams and T.J. Hanratty, "Effect of liquid viscosity on the stratified-slug transition in horizontal pipe flow", *Int. J. Multiphase Flow*, Vol. 15, pp. 877–892 (1989).

D. Barnea, "On the effect of viscosity on stability of stratified gas liquid flow — Application to flow pattern transition at various pipe inclination", *Chem. Eng. Sci.*, Vol. 46, pp. 2123–2131 (1991).

D. Barnea and Y. Taitel, "Structural and interfacial stability of multiple solutions for stratified flow", *Int. J. Multiphase Flow*, Vol. 18, pp. 821–830 (1992).

D. Barnea and Y. Taitel, "Kelvin-Helmholtz stability criteria for stratified flow, viscous versus non-viscous (inviscid) approaches", *Int. J. Multiphase flow*, Vol. 19, pp. 639–649 (1993).

D. Barnea and Y. Taitel, "Nonlinear interfacial instability of separated flow", *Chem. Eng. Sci.*, Vol. 49, pp. 2341–2349 (1994a).

D. Barnea and Y. Taitel, "Structural stability of stratified flow — The two fluid model approach", *Chem. Eng. Sci.*, Vol. 49, pp. 3757–3764 (1994b).

N. Brauner and D. Moalem-Maron, "Analysis of stratified/nonstratified transitional boundaries in horizontal gas-liquid flows", *Chem. Eng. Eci.*, Vol. 46, pp. 1849–1859 (1991).

S. L. Cohen and T. J. Hanratty, "Effects of waves at a gas-liquid interface on a turbulent air flow", *J. Fluid Mech.*, Vol. 31, pp. 467–469 (1968).

M.J. Landman, "Non-unique holdup and pressure drop in two-phase stratified inclined pipe flow", *Int. J. Multiphase flow*, Vol. 17, pp. 377–394 (1991).

Y. Taitel and A. E. Dukler, "A model for prediction flow regime transitions in horizontal and near horizontal gas-liquid flow", *AIChE J.*, Vol. 22, pp. 47–55 (1976).

Y. Taitel and D. Barnea, "Modeling of gas liquid flow in pipes", *Encyclopaedia of two-phase heat transfer and flow I, fundamentals and methods*, Vol. 1, World Scientific Publication Co. Pte. Ltd, Singapore, 2016.

Radial Void Fraction Profiles in a Downward Two-Phase Flow: Reconstruction Using the Surface of Revolution Method

Freddy Hernandez-Alvarado, Simon Kleinbart,
Dinesh V. Kalaga, Sanjoy Banerjee
and Masahiro Kawaji*

*City College of New York,
New York, NY 10031, USA*
kawaji@me.ccny.cuny.edu

Down-flow multiphase reactors offer higher gas phase residence times by slowing down the downward motion of the bubbles due to buoyancy. As a result, longer gas–liquid contact times are achieved compared to conventional bubble columns, but the local void fraction profiles are quite different between the down-flow reactors and conventional bubble columns. To better understand the performance of the down-flow multiphase reactors, the radial void fraction profiles have been measured using gamma densitometry which yields chordal void fraction profiles. A new reconstruction method called the surface of revolution (SOR) method has been developed to convert the chordal void fraction distribution to a radial profile and applied to a down-flow multiphase reactor to study its flow characteristics.

1. Introduction

Down-flow bubble columns (Corrona-Arroyo *et al.*, 2015; Lu *et al.*, 1994; Shah *et al.*, 1983) and their modifications involving the method of gas and liquid injection (Briens *et al.*, 1992; Majumder *et al.*, 2006; Mandal *et al.*, 2005; Ohkawa *et al.*, 1985; Wu *et al.*, 2012) have been extensively investigated because of their high gas hold-up and smaller bubble sizes in comparison to the conventional bubble columns.

The design and development of these down-flow bubble column reactors require a good understanding of the interfacial mass transfer phenomena and interfacial area concentration which in turn depends on the bubble size and void fraction. The local void fraction distribution is also an important design parameter which exerts direct and indirect influences on the reactor performance. So the estimation of the mean void fraction and its three-dimensional distribution within the reactor plays a key role in the design of these reactors.

If the bubbles are very small, typically less than 1 mm, it may be difficult to measure void fraction distributions reliably with conventional measurement techniques such as an optical void probe and wire mesh sensor. On the other hand, gamma densitometry based on the attenuation of a gamma ray beam by a two-phase gas–liquid mixture is a widely used non-invasive measurement technique for reliably measuring the chordal line-averaged void fraction. If line-average void fractions are measured at different chordal positions, it is possible to obtain radial void fraction profiles using some conversion methods. In recent decades, several researchers have used gamma densitometry under different modes of bubble column operation together with different reconstruction techniques to estimate the radial void fraction profiles. Veera and Joshi (1999) used Schollenberger's (Schollenberger *et al.*, 1997) method, while Veera *et al.* (2001) used the Abel inversion to reconstruct the radial void fraction profiles from chordal void fraction profiles measured in batch bubble columns Later, Kumar *et al.* (2012) used Schollenberger's (Schollenberger *et al.*, 1997) method to obtain the void fraction profiles for batch and cocurrent up-flow bubble columns.

Both the Schollenberger and Abel methods employ second-order and fourth-order polynomial functions respectively, to fit the radial void fraction profiles. They have been found to be applicable to batch bubble columns; however, it is difficult to fit the radial void fraction profiles in down-flow bubble column reactors to second- or fourth-order polynomials. Thus, another chordal to radial void fraction reconstruction algorithm needs to be developed without using a polynomial function. This short communication reports on a new method of reconstruction, called the surface of revolution (SOR), which converts the chordal gamma densitometry measurements to radial void fraction profiles for downflow bubble columns.

2. Experimental Setup

The experiments were performed using air and water in a 0.10-m diameter and 64-cm tall acrylic column. The air was sparged from the bottom, and

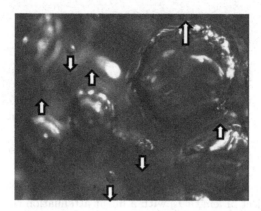

Fig. 1 Typical image taken using a high-speed video camera where the millimeter-sized bubbles rise while the microbubbles move down with the downward liquid flow.

the liquid was injected from the top. A region of high turbulence was maintained in the top section of the column, for generating microbubbles. All the millimeter-sized bubbles injected by the air sparger at the bottom were broken into microbubbles at axial locations higher than $H/D > 6$. Further, due to the small rise velocities, all the microbubbles were entrained by the downward flowing liquid flow and moved downward as shown in Fig. 1. Details of the microbubble generation mechanism can be found in Li (2014). In this microbubble downflow reactor, increasing the liquid flow rate would decrease the residence time of the gas phase. With this configuration, we expect different bubble diameters and, hence, a different void fraction distribution depending on the gas and liquid flow rates.

2.1. *Gamma densitometry*

Gamma-ray densitometry has been used to measure the chordal void fraction at five chordal positions ($x/R = 0$, 0.2, 0.4, 0.6, and 0.8) and at three axial elevations ($H/D = 1.4$, 3.1, and 5.9) under different liquid and gas flow rates as shown in Fig. 2.

Gamma densitometry is a non-invasive and reliable technique for both industrial and lab-scale multiphase reactors (Chavan *et al.*, 2009; Kalaga *et al.*, 2009; Kawaji *et al.*, 1987; Chan *et al.*, 1999). This method is based on attenuation of a gamma ray according to Beer–Lambert's law given by:

$$\frac{I}{I_0} = \exp(-\mu L), \tag{1}$$

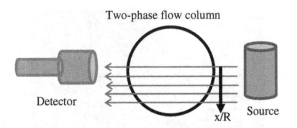

Fig. 2   Schematic of gamma densitometry chordal scans.

where I is the intensity of a mono-energetic gamma beam transmitted through an object of length L, with a linear attenuation coefficient, μ, and I_0 is the incident radiation intensity. The gamma-ray densitometer system used in this work consisted of a 5 mCi Cs-137 source emitting a beam of 662 keV photons and a sodium iodide (NaI) scintillation detector.

The chordal void fraction values were calculated using the following formula,

$$\alpha_{\text{chord}} = \frac{\ln\left(\frac{I_\alpha}{I_1}\right)}{\ln\left(\frac{I_0}{I_1}\right)}, \qquad (2)$$

where I_α, I_1 and I_0 are the gamma-ray intensities detected when the test section has a gas–liquid two-phase mixture, is completely filled with a single-phase liquid and single-phase gas, respectively.

3.   Results and Discussion

After chordal measurements were made with a gamma densitometer, the radial void fraction profiles were reconstructed using the "SOR" method. This method assumes (1) axial symmetry of the void fraction profile and (2) zero void fraction at the wall. It utilizes an axisymmetric body of revolution whose surface represents the radial void fraction profile. An example of a body of revolution is shown in Fig. 3. If a vertical plane intersects the body of revolution at a given chordal location, x_i/R, the dimensionless chord length, $\tilde{L}_i$, is given by,

$$\tilde{L}_i = \frac{L_i}{2R} = \sqrt{1 - \left(\frac{x_i}{R}\right)^2}, \qquad (3a)$$

where R and x_i are the test section radius and chordal location, respectively. If the dimensionless area of the vertical plane under the intersecting curve

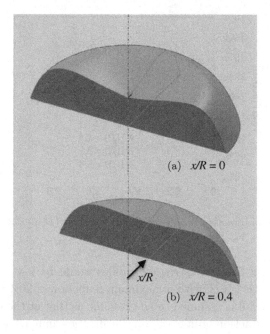

(a) *x/R* = 0

x/R

(b) *x/R* = 0.4

Fig. 3   SOR method.

is, $\tilde{A}_i$, the chordal void fraction would be given by,

$$\alpha_{\text{chord},i} = \tilde{A}_i / \tilde{L}_i. \tag{3b}$$

For example, the vertical plane cutting across the body of revolution at $x/R = 0$ is shown in Fig. 3(a), and at $x/R = 0.4$ in Fig. 3(b), respectively. The shape of the body of revolution is adjusted until all the chordal void fraction values evaluated using Eq. (3b) match the chordal void fraction values obtained from the gamma densitometer.

This method was validated experimentally using a tubular test section surrounded by a water-filled annulus in order to obtain a unique void fraction profile with $\alpha = 0$ at $0.6 < r/R < 1.0$, and a center peaking void fraction, $\alpha > 0$, at $0 < r/R < 0.6$. A detailed explanation of this validation work can be found in Hernández-Alvarado *et al.* (2016).

A sensitivity study was performed for the SOR method in order to check the assumption of zero void fraction at the wall. The void fraction at the wall was assumed to be 0%, 50%, and 100% of the void fraction along the central chord. The results showed that the assumption of non-zero void fraction at the wall would not have a significant effect on the radial void

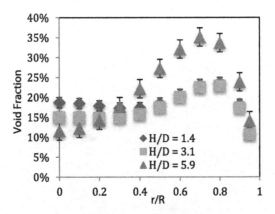

Fig. 4   Radial void fraction profiles at three elevations ($H/D = 1.4$, 3.1, and 5.9) for $V_L = 65$ mm/s, $V_G = 4$ mm/s.

fraction profiles. The reconstructed profiles would be lower at the center and higher at $r/R = 0.8$, with a maximum reduction of 20% when the void fraction at the wall is assumed to be the same as that at the central chord.

Radial void fraction profiles obtained using the gamma densitometer and the SOR method are shown in Fig. 4 for $V_L = 65$ mm/s and $V_G = 4$ mm/s. The void fraction profiles at three axial locations ($H/D = 1.4$, 3.1, and 5.9) are lower at the center, increase until $r/R = 0.8$, and fall to zero at the wall. These profiles are attributed to a recirculation zone existing near the wall where the residence time of the bubbles is increased. Also, at the bottom of the column ($H/D = 1.4$), the local void fraction was found to be higher at the center in comparison with the data obtained at higher elevations which is attributed to the millimeter-sized bubbles injected from the sparger at the bottom. At the top of the column ($H/D = 5.9$), the void fraction profile is lower at the center and higher near the wall at $r/R = 0.8$ compared to the profiles obtained at lower elevations. The overall void fraction is higher and the fall near the wall is even steeper. This is attributed to a possible recirculation of the bubbly mixture from the center to the wall region due to the downward liquid jets. As the large bubbles get broken up by the jets and flow down with the liquid, the radial void fraction profile becomes more uniform at lower elevations.

4.   Conclusions

A gamma-ray densitometer has been used to measure the chordal void fraction profiles in a down-flow bubble column and radial void fraction profiles have been obtained from chordal void fraction measurements using a new SOR method. It utilizes an axisymmetric body of revolution whose surface represents the radial void fraction profile. The reconstructed radial void fraction profile was found to be insensitive to the assumption of zero void fraction at the pipe wall, but a chordal void fraction measurement close to the wall may be necessary to obtain a more accurate radial void fraction profile.

The measurements in a down-flow bubble column showed somewhat different radial void fraction profiles depending on the axial locations and the gas and liquid flow rates. The present results suggest the feasibility of using the SOR method to obtain radial void fraction profiles in test sections where the actual void fraction profile may not be readily fitted by a single polynomial function such as in a down-flow bubble column reactor.

Acknowledgements

The authors are grateful to the Advanced Research Projects Agency-Energy (ARPA-E), U.S. Department of Energy, for the financial support under Award Number DE-AR0000438, and NRC graduate fellowship grants: NRC-27-10-1120 and NRC-HQ-12-G-38-0. Also a special thanks goes to Mr. Jorge Pulido of LanzaTech for his help in the design and construction of the experimental apparatus.

References

1. M. A. Corrona-Arroyo, A. Lopez-Valdivieso, J. S. Laskowski and A. Encinas-Oropesa, "Effect of frothers and dodecylamine on bubble size and gas holdup in a downflow column", *Miner. Eng.*, Vol. 81, pp. 109–115, (2015).
2. X. X. Lu, A. P. Boyes and J. M. Winterbottom, "Operating and hydrodynamic characteristics of a concurrent downflow bubble column reactor", *Chem. Eng. Sci.*, Vol. 49, pp. 5719–5733, (1994).
3. Y. T. Shah, A. A. Kulkarni and J. H. Wieland, "Gas holdup in two- and three-phase downflow bubble columns", *Chem. Eng. J.*, Vol. 26, pp. 95–104, (1983).
4. C. L. Briens, L. X. Huynh, J. F. Large, A. Catros, J. R. Bernard and M. A. Bergougnou, "Hydrodynamics and gas-liquid mass trasfer in a downward venturi-bubble column combination", *Chem. Eng. Sci.*, Vol. 4, pp. 3549–3556, (1992).
5. S. K. Majumder, G. Kundu and D. Mukherjee, "Bubble size distribution and gas-liquid interfacial area in a modified downflow bubble column", *Chem. Eng. J.*, Vol. 122, pp. 1–10, (2006).

6. A. Mandal, G. Kundu and D. Mukherjee, "A comparative study of gas holdup, bubble size distribution and interfacial area in a downflow bubble column", *Chem. Eng. Res. Des.*, Vol. 83, pp. 423–428, (2005).

7. A. Ohkawa, Y. Shiokawa and N. Sakai, "Gas holdup in downflow bubble columns with gas entrainment by a liquid jet", *J. Chem. Eng. Jpn.*, Vol. 18, pp. 172–174, (1985).

8. Y. L. Wu, Q. J. Xiang, H. Li and S. X. Chen, "Study on bubble sizes in a down-flow liquid jet gas pump", *IOP Conf. Ser.: Earth Environ. Sci.*, Vol. 15, pp. 1–7, (2012).

9. U. P. Veera and J. B. Joshi, "Measurement of gas hold-up profiles by gamma ray tomography: Effect of sparger design and height of dispersion in bubble columns", *Trans. I ChemE*, Vol. 77, pp. 303–317, (1999).

10. K. A. Schollenberger, J. R. Torcynski, D. R. Adkins, T. J. O'Hern and N. B. Jackson, "Gamma-densitometry tomography of gas holdup spatial distribution in industrial-scale bubble column", *Chem. Eng. Sci.*, Vol. 52, pp. 2037–2048, (1997).

11. U. P. Veera, K. L. Kataria and J. B. Joshi, "Gas hold-up profiles in foaming liquids in bubble columns", *Chem. Eng. J.*, Vol. 84, pp. 247–256, (2001).

12. S. Kumar, R. A. Kumar, P. Munshi and A. Khanna, "Gas hold-up in three phase co-current bubble columns", *Procedia Engineering*, Vol. 42, pp. 782–794, (2012).

13. X. Li, "System and method for improved gas dissolution", US Patent Application, US 20140212937 A1 (July 31, 2014).

14. P. V. Chavan, D. V. Kalaga and J. B. Joshi, "Solid–liquid circulating multistage fluidized bed: hydrodynamic study", *Ind. Eng. Chem. Res.*, Vol. 48, pp. 4592–4602, (2009).

15. D. V. Kalaga, A. V. Kulkarni, R. Acharya, U. Kumar, G. Singh and J. B. Joshi, "Some industrial applications of gamma-ray tomography", *J. Taiwan Inst. Chem. E.*, Vol. 40, pp. 602–612, (2009).

16. M. Kawaji, Y. Anoda, H. Nakamura and K. Tasaka, "Phase and velocity distributions and holdup in high pressure steam/water stratified flow in a large diameter horizontal pipe", *Int. J. Multiphase Flow*, Vol. 13(2), pp. 145–159, (1987).

17. A. M. C. Chan, M. Kawaji, H. Nakamura and Y. Kukita, "Experimental study of two-phase pump performance using a full size nuclear reactor pump", *Nuclear Eng. Design*, Vol. 193 (1–2), pp. 159–172, (1999).

18. F. Hernandez-Alvarado, D. V. Kalaga, S. Banerjee and M. Kawaji, "Comparison of gas hold-up profiles in co-current, counter-current and batch bubble column reactors measured using gamma densitometry and surface of revolution method". Paper No. FEDSM2016-1025 in Proc. of ASME 2016 Fluids Engineering Division Summer Meeting, Washington, DC, USA, Vol. 1B, pp. V01BT33A006, 8 pages.

Applying Liquid-Liquid Phase Separation for Enhancing Convective Heat Transfer Rates

A. Ullmann*, I. Lipstein and N. Brauner

*School of Mechanical Engineering,
Tel-Aviv University,
Tel-Aviv 69978, Israel*
*ullmann@tauex.tau.ac.il

The possibility of enhancing forced and free convection heat transfer rates by inducing liquid-liquid phase separation of a two-component liquid system with a lower critical solution temperature (LCST) was demonstrated. The LCST system used was composed of water and triethyamine with a critical composition. Visualization of the phase separation on the surface of a hot vertical plate enables to associate the free-convection heat transfer augmentation to the observed flow phenomena.

1. Introduction

With the rapid development of microelectronics technology, high-power density devices are being used for various applications. In order to maintain a temperature that ensures safe and efficient operation of the equipment, high surface heat flux removal is required. To enable efficient cooling, the focus of the heat transfer community has shifted from air cooling to single and two-phase (boiling) liquid cooling. The application of convective boiling for heat removal is constrained by the Critical Heat Flux. Above that heat flux, the surface is covered by vapor (dry-out), leading to a very large increase in the surface temperature. This problem is crucial in microchannels, when the size of the bubble reaches the channel diameter already before its detachment and earlier dry-out occurs. Additionally, the fast growth of elongated bubbles results in instabilities when operating in parallel channels. When considering the combined effects of heat removal, operating pressure losses, operational stability, and device fabrication efforts,

123

liquid flow appears to be the most promising concept for future development of a range of electronic system (Collier and Thome, 1994; Kandlikar, 2004, 2012).

The possibility of using a phase separation of partially miscible liquid–liquid systems to enhance the single-phase heat transfer rates has been examined. The liquid–liquid systems used are partially miscible solvent systems with a critical solution temperature (CST). Such systems can alter from a state of a single liquid phase to a state of two separated liquid phases by a small change of temperature. In liquid–liquid phase separation, density differences are much lower than in gas–liquid systems. However, during the intermediate, non-equilibrium stages of phase separation, the chemical potential gradients are responsible for the so-called Korteweg capillary forces that result in self-propulsion of droplets. Inducing liquid–liquid phase separation at a cooled (hot) surface would, in this case, result in drop detachment, and consequently in inflow of fresh, cold liquid into the thermal boundary layer. Similar to boiling, this pumping mechanism increases the rate of heat removal from the surface. A numerical tool, based on the diffuse interface approach, was developed and used for simulating the separation process and to obtain the transient concentration and temperature fields. The 2D simulation enables the analysis of the evolving velocity field induced by the non-equilibrium Korteweg force (Segal *et al.*, 2012).

Temperature-induced phase separation is encountered in solvent systems that possess either an upper critical solution temperature (UCST) or a lower critical solution temperature (LCST). In UCST systems, the transition from a single phase to two phases is brought about by reducing the temperature, whereas in LCST systems, phase separation occurs with increasing the temperature. The boundary between the complete miscibility of the system and the region where the system separates into two-phases is given by the coexistence (binodal) curve. This curve provides the equilibrium compositions of the two separated phases as a function of temperature (see Fig. 1). For cooling purposes, heat sink is required, whereby a LCST system should be used, while for heating applications, USCT systems should be employed.

2. Forced Convection in Pipe Flow

The feasibility of enhancing convective heat transfer for cooling applications was demonstrated in a series of experiments conducted in a 4 mm I.D. stainless steel pipe using a water–triethylamine (TEA) mixture with an LCST of 18.2°C. The pipe was immersed in a hot water tank to maintain a constant

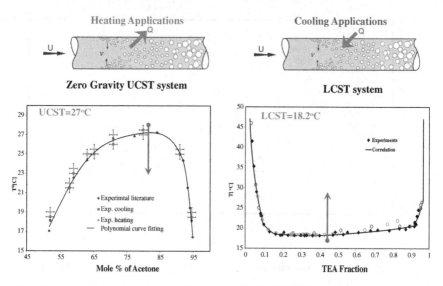

Fig. 1   Coexistence curve of the water–TEA LCST system and acetone-hexadecane UCST system.

wall temperature above the CST. T-type thermocouples were used to measure the inlet and outlet temperatures and the tube inner surface temperature at several locations (spaced evenly). The pressure drop between the outlet and inlet was measured by Validyne pressure transducer. All the experiments were conducted with laminar flow in the pipe, with Reynolds number varying in the range of Re = 120–1100. The Prandtl number of the solvent mixture is typically in the range of 12–17. The reported results correspond to steady state conditions, which are typically achieved after at least five liquid volume replacements in the pipe. More details on the experimental setup and procedure are given in Ullmann *et al.* (2014b).

In order to calculate the total heat transfer in the pipe test section, the non-ideality of a partially miscible solvent system and the associated heat of mixing should be taken into account. The combined effect of results in an effective heat capacity with maxima values at the vicinity of the CST (Fig. 2). The effective heat capacity is used to obtain the total heat absorbed in the flowing liquid as follows:

$$\dot{Q} = \dot{m} \int_{T_{\text{in}}}^{T_{\text{out}}} Cp(T)dT, \quad \dot{m} = \rho_{\text{mix}}\dot{V}. \tag{1}$$

The density of the solvent mixture ρ_{mix} was calculated based on the overall composition and the temperature at the volumetric flow rate ($\dot{V}$)

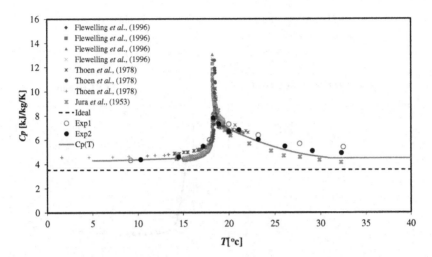

Fig. 2　Effective heat capacity of the water–TEA system. In-house experiments and literature data.

measurement point. The average heat transfer coefficient, h, is defined by

$$h = \frac{\dot{Q}}{A_{\text{in}} \Delta T_{\text{lm}}}; \quad \Delta T_{\text{lm}} = \frac{(T_{\text{out}} - T_w) - (T_{\text{in}} - T_w)}{\ln \frac{(T_{\text{out}} - T_w)}{(T_{\text{in}} - T_w)}}, \quad A_{\text{in}} = \pi D L, \tag{2}$$

where T_w is the tube inner (constant) surface temperature, ΔT_{lm} is the log-mean temperature difference, A_{in} is the internal surface area of the test section, D is the tube inner diameter, and L is the length of the test section.

The experimental setup and calculation procedure for obtaining the convective heat transfer coefficient were tested and validated by conducting control experiments using water flow and single-phase flow of the CST mixtures (where the inlet and outlet temperatures are below the miscibility curve). The results were compared to the values predicted by the well-established correlation proposed by Hausen (1943) for the convective heat transfer in the thermally developing region of the pipe (the flow in the test section is assumed to be fully developed laminar pipe flow) with constant wall temperature:

$$\text{Nu}_D = 3.66 + \frac{0.0668 \, \text{Gz}}{1 + 0.04 \, \text{Gz}^{2/3}}, \tag{3}$$

where $\text{Nu}_D = h D / k$, $\text{Gz} = (D/L)\text{Pe}$, $\text{Pe} = (Dv/\alpha)$ (k and α are the thermal conductivity and diffusivity respectively and v is the average velocity).

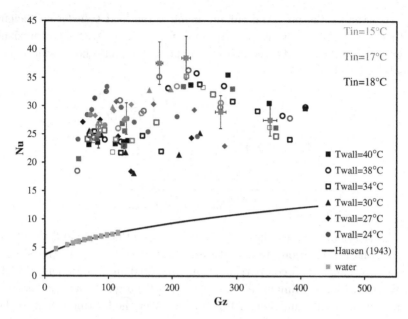

Fig. 3 Experimental results for forced convection Nu number obtained with phase separation of the LCST water–TEA system for various T_w and $T_{in} = 18°C$ ($T_{out} > T_{cp}$ see Fig. 1) — Comparison with the Nu number obtained without phase separation.

The validity of this correlation (Eq. (3)) to predict the single-phase heat transfer coefficient was confirmed. Therefore, for all practical purposes, this correlation can be used as a reference to evaluate the extent of heat transfer enhancement induced by phase separation.

Figure 3 shows the experimental dimensionless heat transfer coefficient ($Nu = hD/k$) obtained in laminar flow with phase separation, in comparison with the values obtained with the same flow rate (same Gz number) without phase separation (depicted in the figure by the single-phase curve). The advantage of inducing phase separation is clearly demonstrated in this figure. The results indicate that phase separation can enhance convective heat transfer coefficients by 50–350% (augmentation factor, $AF = 1.5$–4.5). It is important to note that the augmentation factor represents the ratio of the Nusselt numbers obtained with phase separation to that obtained with single phase flow of the tested liquid system. Therefore, it represents also the Nusselt number augmentation relative to heat transfer rates obtained with any other laminar single-phase flow with the same Graetz number.

To estimate the potential enhancement of the heat transfer coefficient with a water–TEA solvent, a correlation for the heat transfer coefficient augmentation factor, AF (compared to the single-phase flow value, h_{sp}), has been developed:

$$AF = \frac{h}{h_{sp}} = 0.487 \cdot \text{Gz}^{-0.053} \cdot \text{Ja}^{-1.22} \left(\frac{T_{out} - T_{in}}{T_w - T_{cp}} \text{Gz} \right)^{0.664}$$

$$\times \left(\frac{T_{out} - T_{cp}}{T_w - T_{cp}} \right)^{0.475}, \tag{4}$$

$$\text{Ja} = \frac{Cp_{SP}}{\frac{1}{T_{out} - T_{in}} \left[\int_{T_{in}}^{T_{out}} Cp(T)dT \right] - Cp_{SP}},$$

T_{in}, T_{out}, T_w and T_{cp} are the liquid inlet, liquid outlet, wall, and the cloud point (phase transition) temperatures, respectively; Cp_{SP} is the single-phase heat capacity. A comparison of the predicted and experimental AF is depicted in Fig. 4. Ullmann *et al.* (2014a, 2014b) argued that the last two terms on the right-hand side of Eq. (4) represent the heating rate and the depth (in terms of temperature) of the penetration into the unstable region

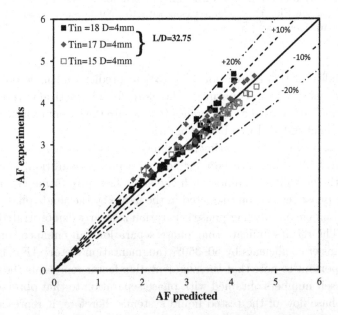

Fig. 4   A comparison of the correlation (Eq. (4)) with the experimental results for the heat transfer amplifications for convective heat transfer with phase separation.

of the coexistence curve. This agrees with previous findings that increasing these two parameters results in higher growth and movement of the separating domains/droplets (Ullmann *et al.*, 2008).

As water is commonly considered to be the preferred choice of working fluid due to its superior heat transport properties, the enhancement of the heat transfer coefficient obtained with the LCST water–TEA mixture, as compared to that obtained with pure water, is of particular interest. Although the thermal conductivity of water is higher than that of the solvent system (by a factor of about 1.7), the heat transfer coefficient obtained with the phase-transition induced heat transfer of the LCST system is between 1.5 and 3 times higher than that of water for the same volumetric flow (Fig. 5).

The ongoing research efforts by the heat transfer community which, target enhancing the single-phase convective heat transfer rates in cooling systems, include various techniques, such as microchannel flows, micropins, sprays, and jet impingement (e.g. Kandlikar and Bapat, 2007; Ebadian and Lin, 2011). In these techniques, microstructures are used to enhance the

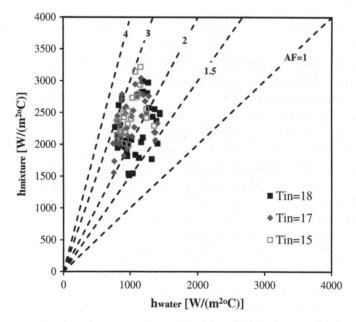

Fig. 5 A comparison of the heat transfer coefficient obtained with the phase transition induced heat transfer of the LCST system to that which would be obtained with water at the same volumetric flow rate.

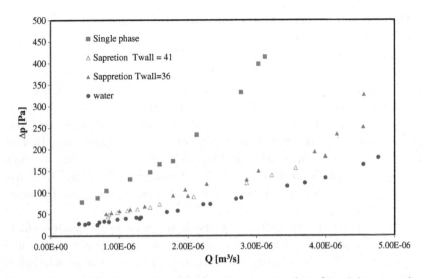

Fig. 6   Pressure drop vs. volumetric flow rate measured in the mini-test section (4 mm I.D.). The pressure drop of the single phase TEA+water is *significantly higher* than that obtained with phase separation.

mixing, and thereby the heat transfer rates. These techniques are generally associated with a penalty in the power pumping demands. With LCST coolants the mixing is achieved by the thermodynamically unstable phase separation. Therefore, in addition to the coolant heat transfer capability, its hydraulic performance is also to be considered for optimum system performance. It is desirable to maximize the heat transfer capability while minimizing the pressure drop/pumping power. Preliminary studies were conducted to examine the effect of phase separation on the pressure drop. Figure 6 shows the pressure gradient obtained with water at the same flow rate and that of the TEA+water system with and without phase separation. The results obtained indicate that the pressure drop does not increase during phase separation (in fact, it decreases with phase separation). Hence, the augmentation of the heat transfer is not associated with increased pressure drop. The somewhat lower pressure drop of water is attributed to the lower viscosity.

The figure of merit (FOM), a commonly used measure for heat transfer performance of coolants, is defined as the ratio between the heat transfer removal rate, $\dot{Q}$, and the pumping power, P, according to

$$\text{FOM} = \dot{Q}/P \quad \text{and} \quad P = \Delta P \dot{V} \tag{5}$$

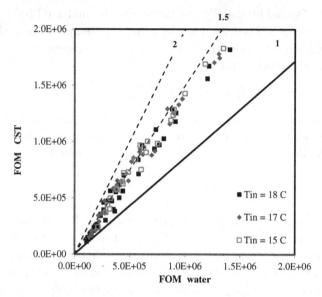

Fig. 7   Results of FOM demonstrating an up to 1.5 fold increase in the FOM of the LCST (undergoing phase separation) compared to water at the same flow rate in the pipe.

Superior coolants are associated with high values of the FOM. Comparisons of the FOM obtained for the LCST mixture (TEA+water) system with that of water at the same flow rate are depicted in Fig. 7. The values shown are based on the heat transfer and pressure drop measurements of the LCST coolant and water. The benefit of the LCST coolant is clearly demonstrated in the figure.

3.   Free Convection Heat Transfer

The water–TEA system was also used to explore the free convection heat transfer phenomena from a vertical plate during phase separation. The experimental setup used is similar to the one described in Ullmann *et al.* (2014a). To this aim, a plate is immersed in a tank filled with the cold (and single phase) CST mixture (with a critical composition). The plate is a square-shaped stainless steel (316-L) plate, 80 mm × 80 mm and 15 mm wide, that is heated by hot water. A temperature-controlled bath and a pump are used for circulating heating water through the plate. T-type thermocouples are used for measuring the inlet and outlet temperatures of the heating water and the temperature of the plate outer surface (at five locations). Two additional T-type thermocouples and a PT100 sensor are used to measure and control the solvent mixture temperature in the tank.

The experimental heat transfer to the plate is calculated by an energy balance on the heating water

$$\dot{Q} = [\rho C p \dot{V} (T_{\text{in}} - T_{\text{out}})]_{\text{water}}, \tag{6}$$

where $\dot{V}$ is the water volumetric flow rate, and T_{in} and T_{out} are the inlet and outlet temperatures of the heating water, respectively. The overall heat transfer coefficient is obtained using the following expression:

$$h = \frac{\dot{Q}}{A_{\text{vert}}(T_w - T_b)}, \tag{7}$$

where A_{vert} is the area of the four vertical surfaces, T_w is the surface (average) temperature, and T_b is the far-field temperature of the CST mixture in the tank. The experimental heat transfer coefficient was compared with that predicted by the well-established correlation for free-convection heat transfer from a vertical plate with a constant surface temperature proposed by Churchill and Chu (1975)

$$\text{Nu}_L = 0.59 \text{Ra}_L^{0.25} \text{ for } 10^4 < \text{Ra}_L < 10^9, \quad \text{Ra}_L = \frac{g\Delta\rho L^3}{\mu\alpha}. \tag{8}$$

However, a better representation of the data in our experimental setup without phase separation is provided by the following correlation:

$$\text{Nu}_{sp} = 0.11 \text{Ra}_L^{0.33}. \tag{9}$$

Figure 8 shows a visualization of flow near the vertical heated plate with and without phase separation. Compared to the single-phase free convection (Fig. 8b), the flow with phase separation (Fig. 8a) is much more "stormy" with apparent irregular interfacial waves. Without phase separation, the driving force for free convection is the density difference due to the thermal expansion $(\beta\Delta T)$. In this case, the heated single-phase fluid is lighter than the bulk density, and therefore it flows in the upward direction. In the case of phase separation, the density difference is also affected by the different compositions of the separated phases. In this case, the bulk density is between the densities of the separated light and heavy phases. This results in more complicated flow phenomena, where the heavy and the light phases flow in opposite directions (i.e. downward and upward, respectively).

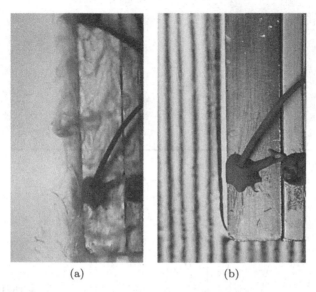

(a)   (b)

Fig. 8   Visualization of free convection from a plate during phase separation (a), and without phase separation (b). Note that the strips in the background were added to enable visualization of the single-phase flow.

Figure 9 shows the experimental dimensionless heat transfer coefficient (Nu = hL/k) obtained with phase separation as a function of the temperature difference between the wall and the liquid bulk ($\Delta T = T_w - T_b$). It is compared with the values obtained without phase separation (depicted in the figure by the single phase curve, Eq. (9)). The advantage of inducing phase separation is clearly demonstrated in this figure. Note that the heat transfer augmentation due to the phase separation is taking place only when the wall temperature is higher than the phase separation temperature, T_{cp}. Accordingly, the ΔT for the inception of the heat transfer augmentation is larger for lower T_b values. Figure 9 also shows a clear linkage between the increased heat transfer rates and the extent of interfacial irregularities and the flow turbulence.

Due to the complicated nature and the opposite directions of the flow of the two separating phases, the film model approach (that was used in Ullmann *et al.* (2014a) for UCST system) seems to be too simplistic for representing the pertinent transfer phenomena. Yet, an empirical correlation is introduced for representing the heat transfer augmentation during

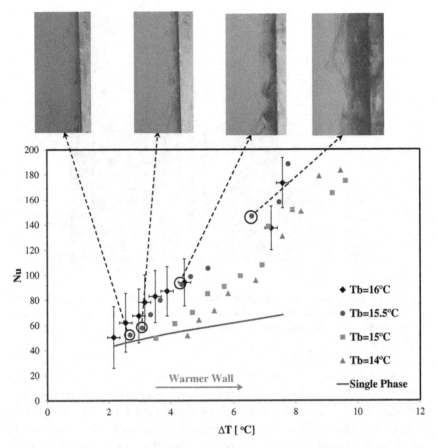

Fig. 9 Free convection from a heated vertical plate during phase separation in comparison with the free convection correlation (Eq. (9)). The flow visualization shows a clear linkage between the increased heat transfer rates and the extent of interfacial irregularities and the flow turbulence.

liquid–liquid phase separation in terms of the quenching depth $(T_w - T_{cp})$

$$AF = \frac{\mathrm{Nu}}{\mathrm{Nu}_{\mathrm{sp}}} = \frac{h}{h_{\mathrm{sp}}} = 0.17\mathrm{Ra}_L^{0.12} \left(\frac{T_w - T_{cp}}{T_{cp} - T_b} \right)^{0.16}, \qquad (10)$$

where h_{sp} is that obtained by Eq. (9) (i.e., free convection without phase separation) for the same Ra_L (which is based on the density difference due to the thermal expansion $\beta \Delta T$, where the expansion coefficient is that of the single-phase mixture). As can be seen in Fig. 10, correlation (10)

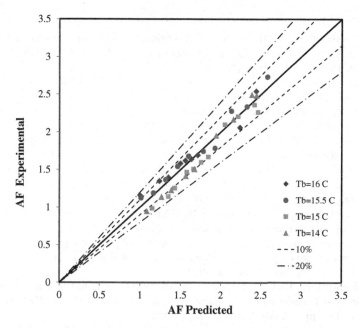

Fig. 10   Free convection augmentation during phase separation of LCST system, comparisons of data with the prediction by Eq. (10).

satisfactory represents the current experimental values (within ±20%), indicating that heat transfer augmentation of up to about 150% can be reached with phase separation of the LCST system.

4.   Summary

Heat transfer rates obtained during phase separation of a two-component liquid system with an (LCST), composed of water and TEA, were studied. Both convective heat transfer rates in small diameter pipe and over a vertical (hot) plate were tested. It was found that phase separation can significantly enhance the forced convection heat transfer in small diameter pipes (by a factor of up to 4.5), and the heat transfer coefficient is by a factor of up to 3 larger than that of water with the same volumetric flow rate. Using the same solvent system and inducing phase separation on a hot vertical plate result in augmentation of the free-convection heat transfer rates by a factor of up to 2.7. Visualization of the flow field near the plate during the phase separation enables to associate the heat transfer augmentation with the observed flow phenomena.

References

S. W. Churchill and H. H. S. Chu, "Correlating equations for laminar and turbulent free convection from a horizontal cylinder," *Int. J. Heat Mass Transfer*, Vol. 18, pp. 1049–1053, (1975).

J. G. Collier and J. R. Thome, *Convective Boiling and Condensation*," 3rd edn., Oxford, University Press, Oxford (1994).

M. A. Ebadian and C. X. Lin, "A review of high-heat-flux heat removal technologies", *J. Heat Transfer T. ASME*, Vol. 133, No. 11, pp. 1–11, (2011).

A. C. Flewelling, R. J. DeFonseka, N. Khaleeli, J. Partee and D. T. Jacobs, "Heat capacity anomaly near the lower critical consolute point of triethylamine–water", *J. Chem. Phys.*, Vol. 104, pp. 8048–8057, (1996).

H. Hausen, "Darstellung des warmeuberganges in rohren durch verallgemeinerte potenzbeziehungen", *VDI Z.*, Vol. 4, p. 91, (1943).

G. Jura, D. Fraga, G. Maki and J. H. Hildebrand, "Phenomena in the liquid–liquid critical region", *Chemistry*, Vol. 39, pp. 19–23, (1953).

S. G. Kandlikar and A. V. Bapat, "Evaluation of jet impingement, spray and microchannel chip cooling options for high heat flux removal", *Heat Transfer Eng.*, Vol. 28, No. 11, pp. 911–923, (2007).

S. G. Kandlikar, "History, Advances and challenges in liquid flow and flow boiling heat transfer in micro channels — A critical review", *J. Heat Transfer*, Vol. 134, pp. 034001-1–034001-15, (2012).

S. G. Kandlikar, "Heat transfer mechanisms during flow boiling in micro channels", *J. Heat Transfer*, Vol. 126, pp. 8–16, (2004).

V. Segal, A. Ullmann and N. Brauner, "Modeling of phase transition of partially miscible solvent systems: Hydrodynamics and heat transfer phenomena", *Comput. Thermal Sci.*, Vol. 4, No. 5, pp. 387–398, (2012).

J. Thoen, E. Bloemen and W. Van Dael, "Heat capacity of the binary liquid system triethylamine–water near the critical solution point", *J. Chem. Phys.*, Vol. 68, pp. 735–744, (1978).

A. Ullmann, S. Gat, Z. Ludmer and N. Brauner, "Phase separation of partially miscible solvent systems: Flow phenomena and heat and mass transfer applications", *Rev. Chem. Eng.*, Vol. 24, No. (4–5), pp. 159–262, (2008).

A. Ullmann, K. Maevski and N. Brauner, "Enhancement of forced and free convection heat transfer rates by inducing liquid–liquid phase separation of a partially-miscible equal-density binary system", *Int. J. Heat Mass Transfer*, Vol. 70, pp. 363–377, (2014a).

A. Ullmann, I. Lipstein and N. Brauner, "Applying phase separation of a solvent system with a lower critical solution temperature for enhancement of cooling rates by forced and free convection", Paper No. 8983 *Proc. Proc. 15th Int. Heat Transfer Conference*, IHTC-15 (2014b).

Scavenging of Soluble Atmospheric Trace Gases Due to Chemical Absorption by Rain Droplets in Inhomogeneous Atmosphere

Tov Elperin*, Andrew Fominykh and Boris Krasovitov

Department of Mechanical Engineering,
Ben-Gurion University of the Negev
P.O.B. 653, 84105, Israel
*elperin@bgu.ac.il

Atmospheric measurements performed over the last few decades revealed an increase in concentrations of soluble trace gases in the atmosphere. The role of precipitation scavenging of trace gases in atmospheric transport modeling is becoming more significant in view of global warming. Indeed, the rise of atmospheric temperature increases the rate of evaporation from the ocean surface and enhances precipitation.

   Soluble atmospheric gases which are scavenged by rain due to chemical absorption can be divided into two groups. Absorption of NH_3, SO_2, HNO_3 is accompanied by aqueous-phase dissociation reactions of the dissolved species. Modeling of scavenging of these gases is conducted using equation of mass transfer with effective solubility parameter that depends on the dissociation constants and pH. Absorption of Cl_2, N_2O_4, O_3, NO, and NO_2 by water droplets is accompanied by chemical reactions of the first and second orders in a liquid phase. Washout of these gases is modeled using mass transfer equation with a sink of the dissolved components in a liquid phase. Due to large number of admixtures in the atmosphere, scavenging of some gases can be determined by absorption that is accompanied by chemical reaction of the order higher than two. The goal of the present investigation is to determine the rate of soluble chemically active gas scavenging in the atmosphere under the combined effect of precipitation and changing altitudinal temperature profile. Particular attention is given to the analysis of the influence of chemical reactions of the first and higher orders and changing pH on the rate of gas scavenging.

Consider the absorption of soluble trace gases having arbitrary solubility by falling rain droplets from a mixture with inert gas in the inhomogeneous atmosphere and accompanied by the nth-order chemical reaction in a liquid phase. The initial vertical distributions of temperature and soluble trace gas concentration in the atmosphere are assumed to be known. The total mass flux density of the dissolved gas transferred by rain droplets is determined by the following expression:

$$q_c = \phi \, u \, c^{(L)}, \tag{1}$$

where u is the terminal fall velocity of droplets, $c^{(L)}$ is concentration of the dissolved gas in a droplet, and φ — volume fraction of droplets in air. The equation of mass balance for soluble trace gas in the gaseous and liquid phases reads

$$\frac{\partial c}{\partial t} = -\frac{\partial q_c}{\partial z} - \phi k_{chn} (c^{(L)})^n, \tag{2}$$

where c is a total concentration of soluble trace gas in gaseous and liquid phases. Since the concentration of soluble chemically active trace gases in the atmosphere is very low (of the order of 1 ppbv), chemical gas absorption and inhomogeneous concentration distribution in the gaseous phase do not affect temperature distribution in the gaseous and liquid phases (see Elperin et al. 2015a, 2015b). At the same time, the influence of the inhomogeneous temperature distribution in the atmosphere on the rate of chemical gas absorption by falling droplets is significant since the solubility parameter and the constant of chemical reaction are temperature dependent. Altitude and temporal evolution of temperature in the atmosphere under the influence of rain was determined in Elperin et al. (2015a). We employ a 1D model of precipitation scavenging of chemically active soluble gaseous pollutants that is valid for small gradients of temperature and concentration. It is demonstrated that transient altitudinal distribution of concentration under the influence of rain is determined by the partial hyperbolic differential equation of the first order. Taking into account initial conditions, we obtain the solution of this equation in a closed analytical form. In order to illustrate the obtained result, we considered scavenging of chlorine (Cl_2) by rain using analytical approximation for the measured altitudinal distributions of this trace gas and temperature in the atmosphere. The results shown in Fig. 1 were obtained for $T_0 = 283\,K$ and rain intensity $5\,mm \cdot hour^{-1}$. Inspection of Fig. 1 shows that after several hours, chlorine

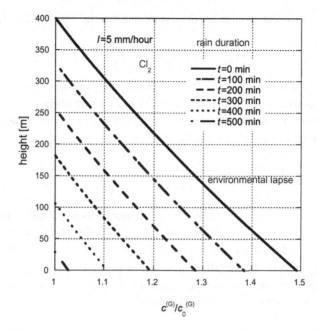

Fig. 1 Evolution of chlorine distribution in the atmosphere caused by rain scavenging.

in the atmosphere is washed out to the concentration equal to the chlorine concentration in the vicinity of the cloud bottom.

Consider absorption scavenging of a soluble gas accompanied by aqueous-phase dissociation reactions. Previous models of gas absorption with aqueous-phase dissociation reactions were developed in the approximation of constant value of pH in the droplet (see, e.g. Elperin *et al.*, 2015a). In this study, we take into account the change of pH in falling rain droplets due to gas absorption. For the dissolution of SO_2 in water, the effective Henry's law coefficient is determined by the following expression:

$$H^*_{SO_2} = H_{SO_2} \cdot (1 + K_{1,SO_2} \cdot 10^{pH}), \qquad (3)$$

where K_{1,SO_2} is a dissociation constant. The expression for pH inside a droplet containing the dissolved sulfur dioxide reads: pH $= -\log\{[H_0^+] + c^{(L)}\}$, where $[H_0^+]$ is concentration of ions inside a cloud droplet without the dissolved sulfur dioxide. Combining equations for total mass flux density of the dissolved gas transferred by rain droplets and mass balance for soluble trace gas in the gaseous and liquid phases with Eq. (3), we obtain the following first-order nonlinear partial differential equation:

$$\frac{\partial c^{(G)}}{\partial t} + (a + bc^{(G)})\frac{\partial c^{(G)}}{\partial z} = 0, \tag{4}$$

where a and b are constants, which depend upon the solubility parameter, dissociation constant, etc. Solution of Eq. (4) for the linear initial distribution of concentration of soluble gas in the atmosphere yields the following expression for the scavenging coefficient:

$$\Lambda = \frac{k'\, a\, (1 + k'\, t\, b) + [c_0^{(G)} + k'\, (z - at)]k'b}{(1 + k'\, t\, b)\, [c_0^{(G)} + k'\, (z - a\, t)]}, \tag{5}$$

where k' is a constant in a linear dependence of concentration from coordinate. Inspection of Fig. 2 shows that Eq. (5) yields the same value of the scavenging coefficient for sulfur dioxide washout by rain as the experimental results reported by Maul (1978) and Martin (1984). Moreover, our theory predicts the same dependence of magnitude of the scavenging coefficient on rain intensity as the experimental results.

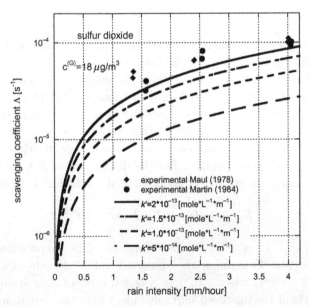

Fig. 2 Comparison of the theoretical predictions with atmospheric measurements of Martin (1984) and Maul (1978) for sulphur dioxide scavenging by rain.

References

T. Elperin, A. Fominykh and B. Krasovitov, "Precipitation scavenging of gaseous pollutants having arbitrary solubility in inhomogeneous atmosphere", *Meteorol. Atmos. Phys.*, Vol. 127, pp. 205–216, (2015a).

T. Elperin, A. Fominykh and B. Krasovitov, "Scavenging of radioactive soluble gases from inhomogeneous atmosphere by evaporating rain droplets", *J. Environ. Radioactiv.*, Vol. 143, pp. 29–39, (2015b).

A. Martin, "Estimated washout coefficients for sulphur dioxide, nitric oxide, nitrogen dioxide and ozone", *Atmos. Environ.*, Vol. 18, pp. 1955–1961, (1984).

P. R. Maul. "Preliminary estimates of the washout coefficient for sulphur dioxide using data from an East Midlands ground level monitoring network", *Atmos. Environ.* Vol. 12, pp. 2515–2517, (1978).

Pin-Fin Two-Phase Microgap Coolers for Concentrating Photovoltaic Arrays

A. Reeser[*,‡], A. Bar-Cohen[*,§], G. Hetsroni[†,¶] and A. Mosyak[†,||]

[*]*Department of Mechanical Engineering,*
University of Maryland, College Park, MD, USA
[†]*Department of Mechanical Engineering,*
Technion-Israel Institute of Technology Haifa, Israel
[‡]*alex.reeser@standardsolar.com*
[§]*abc@umd.edu*
[¶]*hetsroni@tx.technion.ac.il*
[||]*mealbmo@tx.technion.ac.il*

Concentrating photovoltaic (CPV) systems are among the most promising renewable power generation options but will require aggressive thermal management to prevent elevated solar cell temperatures and to achieve the conversion efficiency, reliability, and cost needed to compete with alternative techniques. Two-phase, evaporative cooling of CPV modules has been shown to provide significant advantages relative to single-phase cooling but, to date, the available two-phase data have been insufficient for the design and optimization of such CPV systems.

This contribution will begin with a brief review of CPV technology and the solar cell cooling techniques described in the literature. Attention will then turn to the available correlations for pin-finned microgap coolers and the technology gaps which must be addressed to enable such thermal management for CPV arrays. This will be followed by a detailed description of an experimental study of three pin-finned microgap coolers for CPV systems and the derived heat transfer and pressure drop correlations. The data span a large parametric range, with heat fluxes of 1–$170 \, \mathrm{W/cm^2}$, mass fluxes of 10.7–$1300 \, \mathrm{kg/m^2 \cdot s}$, subcooled (single-phase) flow as well as exit qualities up to 90%, and three heat transfer fluids (water, HFC-134a, HFE-7200). The paper will close with a brief case study of two-phase CPV cooling, demonstrating that the application of this thermal management mode can lead to a highly energy efficient CPV system.

1. Introduction

Multiple-junction solar cells, made from horizontally stacked III–IV semi-conductors, are a most promising alternative to silicon solar cells, with a conversion efficiency that has reached 44.7% with quad-junction cells[a] and is expected to reach even higher values over the coming years. The cell and layers are kept extremely thin — on the order of $8\,\mu m$ for the top layers and $175\,\mu m$ for the bottom substrate layer — to reduce internal series resistances and improve absorption and optical transmission. Each junction is tailored to a specific spectral range with minimal overlap, thereby capturing more of the solar spectrum than silicon and improving efficiency toward the theoretical maximum of 86.8% for an infinite-junction cell (Yastrebova, 2007). Although multijunction III–V solar cells are more expensive than traditional crystalline silicon, the total cell area needed to provide a specified power level can be reduced, due to their inherently higher efficiency and the use of concentration, thus minimizing solar cell material cost. It is expected that concentrating photovoltaics (CPVs), in which the large area of expensive semiconductors is replaced with an equivalent area of relatively low-cost optical reflectors, will lead to considerable cost savings. The power density per unit area of the cell is greatly enhanced by collecting and focusing the light into a small intense beam leading to a reduced cell footprint for comparable power generation. Because of this increased power density and reduced area, the higher cost of the multiple-junction solar cell can then be justified.

The magnification ratio or "suns" of a concentration system is the dimensionless unit by which solar concentrators are compared. It is defined as the ratio of average intensity of the focused light to the standard non-concentrated normal insolation, $1000\,W/m^2$ on the surface of the earth (e.g. 50 suns is $50\,kW/m^2$ of incident power). For high concentration systems of 500 suns or more, the most commonly used optics are point-focus parabolic dish mirrors or Fresnel lenses employed either as multiple, small one-cell systems in series-connected module arrays, or a densely packed "parquet" of cells with one large concentrator. Fresnel lenses function by focusing light via refraction and require a relatively short focal length which can be attained with comparatively less thickness and less material than traditional convex lenses. A parabolic or circular paraboloid dish concentrator works by reflecting all incoming light incident on its surface to a single

---

[a]NREL, "Cell Efficieny Chart," October 2014. [Online]. Available: http://www.nrel.gov/ncpv/images/efficiency_chart.jpg.

focal point, where the receiver containing the cells is located. Parabolic dishes can be scaled up or down in size and have a theoretical concentration limit of 10,000 suns. This factor is lower in practice due to imperfections in the reflecting surface, but 2000 suns or more is attainable (Luque and Hegedus, 2011). Some disadvantages of CPV systems that currently prevent widespread use are: the need for dual-axis tracking systems that add significantly to the system cost and complexity; decreased effectiveness in concentrating diffuse sunlight, which can constitute a large fraction of the incident solar energy in certain locations; and finally, the need for active cooling systems. Despite these limitations, CPV remain very promising for utility scale and high power installations.

Solar cells, like most semiconductor-based electronic devices, are adversely affected by temperature. When the temperature rises, more electrons are excited into the conduction band and, in a PV cell, this has the effect of reducing power conversion efficiency. The relationship of cell efficiency to temperature is commonly expressed as a simple, but useful linear equation which is expected to be quite accurate up to temperatures of about 350°C (Landis, *et al.*, 2005). The efficiency quoted by manufacturers of solar cells is typically at ideal conditions, with the cell operating at 25°C in direct sunlight. Manufacturers will specify a mean value of the temperature coefficient for a large population of cells and a maximum continuous operating temperature, for example, about 100°C for Spectrolab C4MJ cells.[b] It is to be noted that the temperature coefficient is difficult to measure and can vary significantly depending on various parameters such as the type, diameter, thickness, and configuration of the semiconductors used, the spectrum and concentration level of light in which it is being tested, and cell-to-cell manufacturing inconsistencies.

Figure 1 shows a comparison between production silicon, GaAs, and triple junction cells over a 25–100°C range (Green *et al.*, 2012). It can be seen from the figure that the cell type and operating temperature can play an important role in cell efficiency and, hence, performance, especially at increasingly higher temperatures. The operating temperature of PV cells will always be above ambient, without a "cooling solution," due to the need to dissipate the heat generated by the absorbed, but "unconverted," incident sunlight. Cell-to-ambient temperature differences are typically $20-30°$, when PV arrays are exposed to direct sunlight.

---

[b]Spectrolab, [Online]. Available: http://www.spectrolab.com/DataSheets/PV/CPV/ C4MJ_40Percent_ Solar_Cell.pdf.

A. Reeser et al.

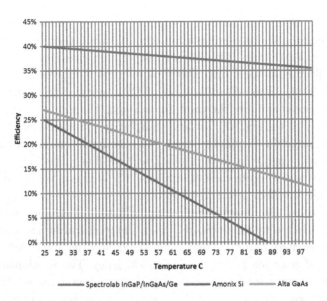

Fig. 1 Comparison of three types of photovoltaic cells. Spectrolab InGaP/InGaAs/Ge triple junction, Amonix single junction Si, and Alta single junction GaAs (Green, *et al.*, 2012).

Despite the decreased conversion efficiency at elevated temperatures, it is unusual for non-concentrating solar cells to be actively cooled, due to the modest amounts of waste heat and the inability to recover the additional expense and parasitic cooling system power consumption of the cooling system. With increased concentration ratios, the loss in efficiency becomes far more significant along with the risk of severe and permanent thermal degradation, necessitating active thermal management for CPV systems.

Verlinden *et al.* (2008) describe Solar Systems' production CPV system, based in Australia and capable of reaching 500 suns. It employs 40% efficient Spectrolab triple-junction cells and uses an improved optical design with multiple reflecting mirrors assembled in a parabolic shape. The dish design allows the system to achieve a geometric concentration of 500 suns. A single array in one dish receiver consists of 64 "modules" and 1500 individual cells with module efficiencies as high as 36.1% and a total rated system output of 36.5 kW per dish. A single-phase liquid cooling system is used (Ho, 2010) to remove the dissipated heat, requiring 950 W, or less than 3% of total system output, to power the cooling system.

Currently, no production CPV systems are cooled using two-phase flow and boiling. Although the technology has promise for CPV cooling, due to the low pumping power requirements and excellent heat transfer rates, the thermofluid transport mechanisms for flow boiling are not yet predictable with the high accuracy needed for solar system design. The only two-phase cooling PV study reported in the literature is by Ho (2010). The author analytically compared single-phase water and two-phase R134a for their high aspect ratio 1 m × 100 mm wide, single-channel cooler under 100 suns. They compared several flow rates, channel heights, and inlet temperatures and their effect on cell efficiency and performance. From their analysis, they concluded that R134a was the superior fluid for two-phase cooling due to its low saturation temperature and low required pumping power.

As CPV concentrations begin to exceed 500 suns, heat fluxes are generated at the cell surface which cannot be easily removed with a $10 - 20°C$ temperature rise, even with the high two-phase heat transfer coefficients. The need for area enhancement becomes more critical with these higher concentration ratios. Pin-fin microgap coolers are an excellent candidate to manage the higher heat fluxes generated by 500+ sun systems.

2.   Available Heat Transfer Correlations

A summary of the best single-phase micropin-fin correlation found in the literature, for heat transfer coefficient and frictional Pressure drop, proposed by Tullius *et al.* (2012), is given in the following section. It was developed for a range of conditions, including various pin fin shapes, sizes, and heat sink materials using water as the working fluid. It was found to have good prediction accuracy of 8–9% MAE for the heat transfer coefficient and 6–9% MAE for the pressure drop (depending on shape of the pin fins). The Tullius *et al.* correlation can be applied from micro- to mini-sized pin-fins, as well as for a large range of heat flux (10–150 W/cm^2), mass flux (60–1000 kg/m^{2}s), and Reynolds number (100–1500). It is to be noted that successful correlation of the data required a distinct geometric factor, C_{Nu} and C_f for each pin-fin shape. The Tullius *et al.* Nusselt number correlation for pin fins is given as (Tullius, *et al.*, 2012):

$$Nu_f = C_{Nu} \left(\frac{S_L}{D_f}\right)^{0.2} \left(\frac{S_t}{D_f}\right)^{0.2} \left(\frac{h_f}{D_f}\right)^{0.25}$$
$$\times \left(1 + \frac{dh}{D_f}\right)^{0.4} Re_f^{0.6} Pr^{0.36} \left(\frac{Pr}{Pr_s}\right)^{0.25}, \tag{1}$$

where the value of C_{Nu} varies with geometry, as shown below

Geometry	Circle	Square	Diamond	Triangle	Ellipse	Hexagon
C_{Nu}	0.08	0.0937	0.036	0.0454	0.0936	0.0752

For frictional pressure drop, Tullius *et al* used a similar correlational form along with a shape multiplier as below (Tullius *et al.*, 2012):

$$f = C_f \left(\frac{S_L}{D_f}\right)^{0.2} \left(\frac{S_t}{D_f}\right)^{0.2} \left(\frac{h_f}{D_f}\right)^{0.18} \left(1 + \frac{dh}{D_f}\right)^{0.2} Re_f^{-0.435}, \qquad (2)$$

where the value of C_f varies with geometry, as shown below

Geometry	Circle	Square	Diamond	Triangle	Ellipse	Hexagon
C_f	2.96	5.28	1.81	2.45	3.44	4.53

The summary of all the two-phase micropin-fin correlations, used in this work for the heat transfer coefficients and frictional pressure drop, is given below. All the two-phase micropin-fin heat transfer correlations found in the literature were developed for highly subcooled inlet conditions and low exit thermodynamic vapor qualities. At the time of this work, no studies were found for saturated or near-saturated inlet conditions nor for high vapor quality flow conditions.

The correlation for heat transfer coefficient by Krishnamurthy and Peles (2008) was developed for high heat flux cooling (20–350 W/cm^2) with a silicon pin-fin microcooler, having circular staggered pin fins of 100 μm diameter. It uses a superposition type model, with the nucleate boiling term removed. The single-phase Nusselt number relation is believed to be valid for Reynolds numbers less than 10^3. The correlation is given in Eq. (3) as follows:

$$h_{tp} = \zeta(\phi^2)^{0.2475} h_{sp}, \qquad (3)$$

$$h_{sp} = \frac{Nu \cdot k_f}{d_{\mathrm{fin}}},$$

$$Nu = 0.76 \left(\frac{S_t}{d}\right)^{0.16} \left(\frac{S_L}{d}\right)^{0.2} \left(\frac{H_{fin}}{d}\right)^{-0.11} Re^{0.33} Pr^{0.333},$$

$$(\phi)^2 = 1 + \frac{0.24}{X_{vv}} + \frac{1}{X_{vv}^2} \quad \zeta = 1,$$

$$X_{vv} = \left[\frac{(\Delta P_f / \Delta Z)_f}{(\Delta P_f / \Delta Z)_v} \right]^{1/2} \quad (\Delta P_f)_f = \frac{fN(G(1-x))^2}{2\rho_f}$$

$$(\Delta P_f)_v = \frac{fN(Gx)^2}{2\rho_v}.$$

In Eq. (3), X_{vv} is the Martinelli parameter, N is the number of pin-fin rows in the flow direction, f is the single-phase friction factor, x is the exit quality, S_t, S_L, and H_{fin} are the transverse length, longitudinal length, and height of the fins respectively.

The Qu and Siu-Ho (2009) correlation was developed for high heat flux cooling (25–$250\,\text{W/cm}^2$) utilizing a square, staggered copper pin-fin array with a subcooled inlet. The model was fitted to Qu and Siu-Ho's original data and requires a subcooling term, in the form of negative inlet quality, in order to obtain proper results. It is, therefore, not applicable to a fully saturated inlet condition, but is presented below for completeness

$$h_{tp} = 1.0 - 12.2 \cdot x_{\text{in}} \exp\left[-(101 \cdot x_{\text{in}} + 29.4) \cdot x_e\right] \cdot 50.44, \qquad (4)$$

where x_{in} is the inlet subcooling and x_e is the local quality.

The heat transfer coefficient model developed by McNeil *et al.* (2010) is for relatively low heat flux cooling (1–$15\,\text{W/cm}^2$) using refrigerant R113 in copper inline pin fins. Following the form of the venerable Chen correlation (David, *et al.*, 2014) and similar to the Krishnamurthy and Peles model, it utilizes superposition, which addresses the nucleate boiling and convective heat transfer mechanisms separately. It is the only micropin-fin correlation in this study that was developed for two-phase inline pin-fin arrays. The correlation is given below.

$$h_{tp} = S \cdot h_{nb} + F \cdot h_{\text{conv}} \qquad (5)$$

In the McNeil *et al.* correlation, the single-phase convective term is given as

$$h_{sp} = \frac{\text{Nu} \cdot k_f}{d_{\text{fin}}},$$

$\text{Nu} = \text{Nu}_r \times F_1 \times F_4$ (F_4 is a row dependent multiplier)

$\text{Nu}_r = a \cdot \text{Re}_b^m \text{Pr}_b^{0.34}$

For $\text{Re} < 300, a = 0.742, m = 0.431$

For $300 < \text{Re} < 2 \times 10^5$, $a = 0.211, m = 0.651$

For Re $> 2 \times 10^5$, $a = 0.116$, $m = 0.7$

$$F_1 = \left(\frac{\mathrm{Pr}_b}{\mathrm{Pr}_w}\right)^{0.26}$$

The nucleate boiling term in the McNeil *et al.* correlation is expressed in terms of the reduced pressure, P_r, as:

$$P_r = \frac{P}{P_{\mathrm{cr}}},$$

$$h_{nb} = 0.945 P_r^{0.17} + 4P_r^{1.2} + 10P_r^{10}(P_{\mathrm{cr}}/1000)^{0.69}(q''/1000)^{0.7}.$$

The Enhancement and Suppression factors for Eq. (5) are given by

$$X_0 = 0.041\left[\frac{\sigma}{g\left(\rho_l - \rho_v\right)}\right]^{1/2} \quad F = \left(\phi_L^2\right)^{0.36},$$

$$X_c = F \cdot h_{\mathrm{conv}}\frac{X_0}{k} \quad (\phi_L)^2 = 1 + \frac{8}{X_{tt}} + \frac{1}{X_{tt}^2},$$

$$S = \frac{1}{X_c}\left(1 - e^{-X_c}\right) \quad X_{tt} = \left(\frac{1-x}{x}\right)^{0.9} * \left(\frac{\rho_v}{\rho_l}\right)^{1/2}\left(\frac{\mu_l}{\mu_v}\right)^{0.1}.$$

3. Description of the Experimental Apparatuses

In the present work, two separate copper micropin-fin arrays of staggered and inline configuration were manufactured on equal overall base areas as well as equal pin width and height, so that performance of the two arrays may be directly compared. In this section, a detailed overview of the testing loop is provided and the experimental procedure used to evaluate the thermofluid performance of these micropin-fin channels is discussed.

Test Loop: The following devices were used in the experiment: the micropin-fin test section, liquid-cooled condenser, liquid reservoir, fluid pump, rotameter, two inline heaters, inlet and outlet pressure transducers, and various E-type thermocouples for reading fluid and test section temperatures. Semi-transparent, high temperature, flexible silicone rubber tubing was used to connect these devices and provided a robust and easily customizable test vehicle for the current setup.

A schematic of the testing loop is shown in Fig. 2. For pressure readings, two separate transducers were used, one at the inlet and one at the outlet, so that inlet and outlet pressures could be measured independently. This was done to enable determination of the liquid subcooling, confirmation of saturated boiling condition, and the vapor quality at the exit, along with

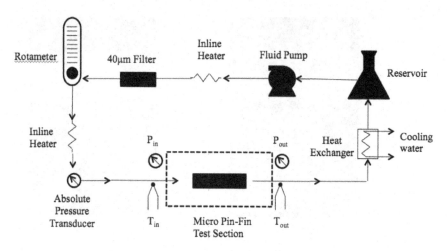

Fig. 2   Micropin-fin testing loop.

temperature readings. Two inline heaters were used to heat the fluid to the desired inlet temperature, with sufficiently low power in each heater to avoid premature boiling or liquid dryout inside the heaters before reaching the test section. A McMaster-Carr 40 μm inline filter was inserted upstream of the test section to prevent contaminants from clogging the micropin-fins. The rotameter is an Omega FL-5000 series flow meter with interchangeable tubes. It was installed with a 305 cc/minute maximum flow rate tube with 0.15 cc/minute measurement markings. Flow readings are measured visually with the metal ball float, and flow rate can be controlled with the integrated valve. The condenser is a flat plate heat exchanger, cooled with forced convection of water. The flow rate of the cooling water was manually controlled to condense the working fluid and lower the working fluid temperature to the desired value before entering the reservoir. The pressure transducers were Setra Systems Model 230 with voltage signals between 0.05 and 5.05 V. The pressure range for each sensor was 0–50 psi and 0–5 psi, for the inlet and outlet sensor, respectively. An absolute pressure transducer was used to verify the inlet pressure reading.

Pin-fin Arrays: Two micropin-fin arrays, a staggered configuration and inline configuration, were fabricated out of copper, using a wire electric discharge machining (EDM) process. The arrays had equivalent base areas of 0.96 cm × 2.88 cm and used identical square pin-fin width and height of 153 μm and 305 μm, respectively, to allow for direct performance comparisons. Due to their orientation, the inline and staggered arrays differ in

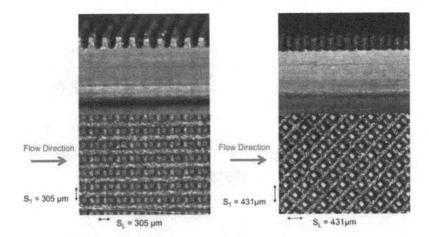

Fig. 3   Inline and staggered pin-fin arrays.

their transverse and longitudinal spacing, with both dimensions equaling 305 μm for the inline array, while both spacings equaled 431 μm for the staggered array. Figure 3 shows a side-by-side visual comparison of these two arrays.

Three, approximately 1 cm^2 square ceramic heaters were soldered on the back of each array, using 63% Sn/37% Pb electronic grade solder paste. Ten small holes were drilled above the heated surface, where thermocouples were inserted to measure the wall temperature of the test section. One polycarbonate (Lexan) housing was manufactured to fully enclose the pin-fin array being tested, while providing insulation from natural convection heat losses during testing. The housing and pin-fin arrays were designed such that easy replacement of test sections could be accomplished, as needed, with no other modification to the testing loop. On top of the housing, a polycarbonate cover was attached and sealed with silicone RTV. Figure 4 is an exploded view of the full assembly.

Test Procedure: The procedure to obtain single-phase data was as follows: the flow rate was set to the desired value using the rotameter. Next, the inline preheaters were turned on and set to a power that would yield the inlet temperature for the tests. The heat exchanger cooling water flow was then turned on. A low initial heating level was applied at the test section, using the power supply. Heat was increased in small increments for each test, and the system was allowed to reach steady state, which typically required 2–3 min, before data readings were gathered. The procedure

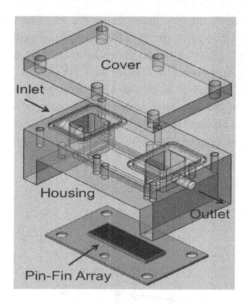

Fig. 4 Exploded view of test section assembly.

to gather two-phase data was similar to the single-phase procedure. After all tests with water were completed, the testing loop was drained of all fluid and allowed to dry for several days. Afterward, the testing loop was charged with HFE-7200 and similar testing procedures to water were performed. Two runs of each mass flux for staggered and inline arrays were run with repeatability in the range of 2–5%.

Additional experiments were performed using a similar micropin-fin heat sink designed by David *et al.* (2014). The heat sink was fabricated of copper, again using the electric discharge machining method. Silver over nickel plating was applied to the heat sink to allow soldering of the resistor to the heat sink base with overall dimensions of $1 \times 1 \, \text{cm}^2$. The pin-fins are $0.35 \times 0.35 \, \text{mm}$ in diameter, $1 \, \text{mm}$ in height, and have a pitch of $0.45 \, \text{mm}$. The dimensions were chosen based on typical microscale devices described in the literature. To minimize the pressure drop, the pin-fins were designed in staggered arrangement, with pin-fin corner normal to the flow direction.

4. Single-phase Micropin-fin Experiments

Single-phase experiments were performed with deionized water and HFE-7200 in both the staggered and inline arrays. The tests established a baseline to which the available correlation could be compared, as well as to gauge the

relative cooling performance enhancement for the two-phase flow boiling
experiments. Inlet temperature for all single-phase experiments was held
constant at 30°C. Plots of single-phase average heat transfer coefficient
vs. heat flux for both deionized water and HFE-7200, in the inline and
staggered arrays, are given in this section. Results were corrected for fin
efficiency, and the average heat transfer coefficient is based on the total
wetted area of the channel. In Fig. 5, the results for deionized water in the
inline array and staggered array are shown for four different mass fluxes

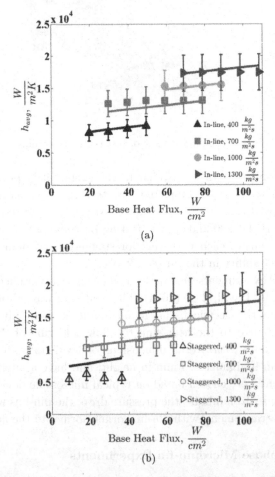

Fig. 5   Modified Tullius *et al.* prediction for single-phase water in the inline array
(a) using a 0.0495 shape factor and the staggered array (b) using a 0.0413 shape
factor.

from 400 to 1300 kg/m²s (calculated using the open area at the entrance to the test section) and heat fluxes in the range of 10–110 W/cm². A detailed error analysis suggests that the measured values are within the ±16% error bars shown in the figures.

It can be seen that while the average heat transfer coefficient is almost independent of heat flux (zero slope) for a constant flow rate, a 3× increase in mass flux will cause the average heat transfer coefficient to increase by about 2×. Additionally, it can be seen that the heat transfer coefficients for the inline and staggered arrays are similar in magnitude for equivalent mass fluxes, with the inline array slightly better, except for the highest mass flux of 1300 kg/m²s.

The Tullius *et al.* correlation prediction, as published, for the inline and staggered arrays provides square and diamond shape factor multipliers of 0.0937 and 0.036, respectively. Using these values, significant mean average errors (MAE) of 87.52% are found between the correlation and the inline array data. However, since the correlation was developed exclusively from staggered array data, the UMD inline array is outside the parametric range of the Tullius correlation. A far lower MAE of 16.09% was found for the staggered array. Since the current data are within the parametric range of Tullius *et al.*, higher accuracy is anticipated for the staggered configuration, but the agreement between the data and the correlation is still not as good as the MAE of 9% reported by Tullius *et al.* Interestingly the accuracy of the correlation can be improved to MAE of 3.48% for the inline array by altering the shape factor to 0.0495, and to an MAE of 12.07% for the staggered array by altering the shape factor to 0.0413. The comparison of the modified Tullius *et al.* correlations and the UMD single-phase data is shown in Fig 5. It is to be noted that the selected shape factors are within range of the published shape factors values, as stated above.

The results for HFE-7200 average heat transfer coefficient versus heat flux for the inline and staggered arrays are shown in Fig. 6 for three different mass fluxes from 200 to 600 kg/m²s, with an expected experimental discrepancy of ±16%. The magnitude of the heat transfer coefficients is lower than for water due to HFE-7200's relatively poor conductive and convective thermal properties. It may also be noted that the HFE data display a somewhat stronger dependence on heat flux than seen in the single-phase water data. At the same mass flux, the staggered array was found to provide 30–50% higher heat transfer coefficients than the inline arrays, for the two highest mass fluxes.

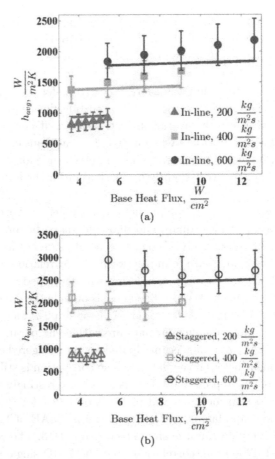

Fig. 6   Modified Tullius *et al.* prediction for single-phase HFE-7200 in the inline array (a) using a 0.054 shape factor and the staggered array (b) using a 0.065 shape factor.

Similar to water, the inline array displays a 2× improvement in the average heat transfer coefficient for a 3× increase in mass flux and is relatively insensitive to the applied heat flux. However, for the staggered array, a nearly 3× improvement in the average heat transfer coefficient occurs for a 3× increase in mass flux and the heat transfer coefficient appears to display a more complex dependence on heat flux than seen with water. Relatively high MAE values were found between the Tullius *et al.* correlation and data, reaching 70.47% overall for the inline array and 36.49% for the staggered array. As seen in Fig. 6, displaying the measured UMD data and the modified Tullius *et al.* correlation, the accuracy of the correlation can be improved to 9.28% for the inline array by setting the shape factor to 0.054, and to 23.35% for the staggered array by setting the shape factor to 0.065. While these shape factors differ from the values proposed by Tullius *et al.* for the square and diamond shaped pins, with 0.0937 and 0.036, respectively, they do fall within the stated range of these shape factors.

For the additional heat sink tested by David *et al.* in Hetsroni's laboratory at the Technion, David *et al.* (2014) experiments were performed in the range of heat flux, q', from 5.2 to 24.7 W/cm^2 and mass flux, G, from 10.7 to 39.1 kg/m^{2}s. A thermal high-speed imaging radiometer was utilized to study the temperature field on the electrical heater, focusing on temperature non-uniformity (on the heated surface) under conditions of convective heat transfer of water in a microchannel without pin-fins.

The temperature distribution on the heated wall depends on the material and design of the test module, flow rate in the microchannel, heat flux, and type of working fluid. The infrared image and histogram of the temperature distribution on the heated side of the test module are shown in Figs. 7(a) and (b) for G equal to 34.5 kg/ m^{2}s and a heat flux, q', of 16.0 W/cm^2. The flow is from the left to the right. The area of the heater (the marked square 2) is clearly shown, and the thermal image analysis is restricted to this square area of 9.5 × 9.5 mm^2. As can be seen from the histogram, the temperature of the resistor was mainly concentrated around 50°C and did not deviate much from this value. Figures 8(a) and (b) show temperature distribution (line 1) on the center of the heated surface in the flow direction at Gm = 22.2 kg/m^{2}s, $q'' = 11.6$ W/cm^2, and Gm = 34.5 kg/m^{2}s, $q'' = 16.0$ W/cm^2. Though the graphs do not display a constant temperature, they have temperature fluctuations with only small values of standard deviation. It may be concluded that the temperature distribution across the resistor in the direction of flow is nearly uniform. Comparison of temperature distribution between pin-fin and smooth microchannels at the same mass flux and heat flux conditions is shown in Figs. 9(a) and (b). The heat transfer coefficient in a smooth microchannel was calculated using the theoretical value of the Nusselt number, i.e. Nu = 4.36. It does not depend on mass flux.

In the pin-fin heat sink, the heat transfer coefficient increases along the flow direction. When a fluid flows across pin-fins, centrifugal forces cause secondary fluid motion, which gives rise to increased heat transfer rates. At very short distances from the start of heat transfer zone, the thermal boundary layer near the base of heat sink is too thin to be affected by secondary

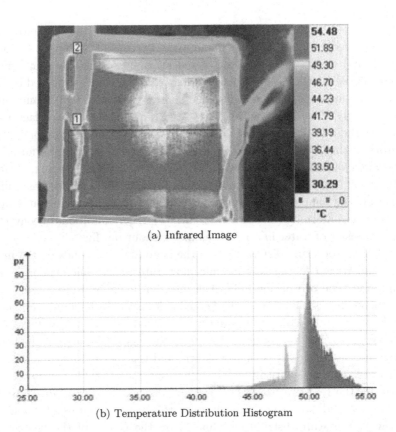

(a) Infrared Image

(b) Temperature Distribution Histogram

Fig. 7   Temperature distribution (T_W) on the whole heated surface of the test module (marked as the square 2) at $G = 34.5\,\text{kg/m}^2\text{s}$, $q = 16.0\,\text{W/cm}^2$.

flow field. Therefore, near the inlet pin-fins offer only a small advantage over a straight microchannel. For greater axial distances, the enhancement factor increases. It should be noted that the smooth microchannels follow an established relationship in which the temperature of the fluid changes linearly in the flow direction and that the temperature of the item being cooled follows this same linear rate of change but at higher temperatures. From Figs. 9(a) and (b), one can see that the use of a microchannel with micro-pin-fins results in more uniform temperature distribution as compared to smooth microchannels. *Temperature measurement results*: Figure 10 depicts the average temperature, T_W, of the heater, as measured with the radiometer. For a given heat flux, q'', T_W decreases with increasing mass flux, G. At given mass flux, T_W increases linearly with increasing the heat flux. The

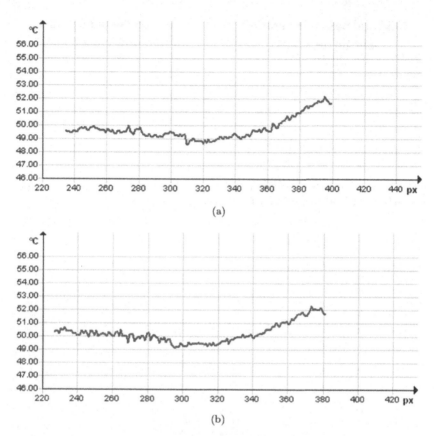

Fig. 8   Temperature distribution (TW) on the whole heated surface (marked as the square 2) at $G = 34.5\,\mathrm{kg/m^2s}$, $q = 16.0\,\mathrm{W/cm^2}$.

overall trend in the measured wall temperature is typical of a single-phase heat transfer system. It should be stressed that local temperatures of the heater in the present experiments are very close to T_W.

One drawback of a microchannel heat sink with single-phase cooling is a relatively high axial temperature rise along the microchannels compared to that for traditional heat sink designs. In the microchannel heat sink, the large amount of heat generated in a high CPV system is removed from the PV array by a relatively low coolant flow rate. Large axial temperature rises produce thermal stresses in electronic elements and packages due to the differences in the coefficient of thermal expansion, thus undermining the device's reliability. This temperature rise may be accompanied by a

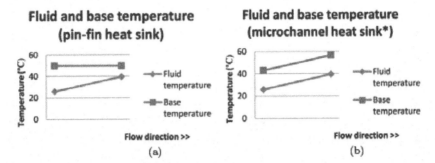

Fig. 9   Comparison between temperatures on the heated wall, $G = 34.5\,\mathrm{kg/m^2 s}$, $q = 16.0\,\mathrm{W/cm^2}$ (a) microchannel with pin-fins, (b) smooth microchannel.

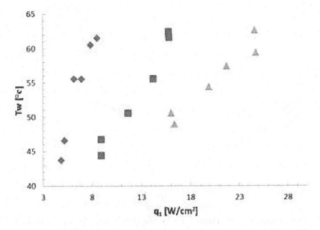

Fig. 10   Variation of the average wall temperature measured by IR with input heat flux. $\Diamond - G = 11.6\,\mathrm{kg/m^2 s}$, $\Box - G = 23.8\,\mathrm{kg/m^2 s}$, $\triangle - G = 36.8\,\mathrm{kg/m^2 s}$.

complex pattern of spatial temperature variations that can produce potentially destructive thermal stresses along the interface between the solar cell and the heat sink. This is one of the key justifications for seeking a nearly isothermal heat sink. Furthermore, a large temperature gradient is undesirable for the performance of the PV cells since many electronic parameters are adversely affected by substantial temperature gradients. For instance, electrical–thermal instability occurs within a high temperature region, because the base elements of PV cells have a switching time that decrease with increasing temperature. In the present experiments (with a range of heat flux from 5.2 to 24.7 W/cm² and mass flow rate from 0.062 to 0.23 g/s) involving very small coolant flow rates and using a pin-fin microheat sink, the temperature distribution on the heater recorded by an IR camera showed a standard deviation from the average temperature of the heated wall of 0.3–1.9 K and that the maximum difference of the wall temperature between different points does not exceed 3–5 K.

5. Two-phase Micropin-fin Experiments

Two phase flow boiling experiments were performed with deionized water and HFE-7200 in both the staggered and inline arrays. The goals of the experiments were to evaluate the cooling enhancement that flow boiling could provide over the single-phase baseline and to determine the accuracy of the existing two-phase correlations for prediction of heat transfer coefficients at high exit qualities. The results would support the determination of the best pin-fin configuration for energy efficient cooling at the high heat fluxes that are encountered in a CPV array. In the following sections, the two-phase cooling experiments are described along with a comparison of the results with those available in the existing literature. Plots of two-phase average heat transfer coefficient vs. exit quality for both deionized water and HFE-7200, in the inline and staggered arrays, are given in this section. Results were corrected for fin efficiency, and the average heat transfer coefficient is based on the total wetted area of the channel.

Water entered the test sections at about 95°C, keeping the subcooling low so as to subsequently allow exit qualities to be as high as possible, while keeping surface temperatures below 140°C to prevent thermal destruction of the test apparatus or any of the components. Figure 11 is a plot of the two-phase water average heat transfer coefficient vs. exit quality for the inline and staggered arrays for four different mass fluxes from 400 to 1300 kg/m^{2}s and heat flux from 27 to 118 W/cm^2. The expected experimental uncertainty of ±16% is indicated by the error bars.

Distinct trends can be observed for each mass flux, with the heat transfer coefficient at the same exit quality increasing with mass flux. As the exit quality increases, the average heat transfer coefficient also monotonically increases, with all data points better than the respective single-phase asymptote marked on the y-axis as 0% exit quality. It is also important to note that the inline and staggered data points nearly coincide over the entire range of qualities shown here, implying that neither the inline nor staggered array is significantly better than the other in terms of cooling performance.

Comparison of the current water data with the available two-phase correlations outlined in Sec. 1 reveal the large differences in the trend and

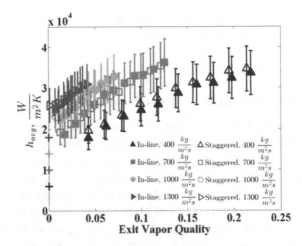

Fig. 11   Average two-phase heat transfer coefficient versus exit quality for water in the staggered and inline pin fin arrays. Error bars indicate 16%. "+" marks indicate single phase asymptotes.

magnitude of the predicted heat transfer coefficients among these correlations. While the heat transfer coefficients are observed to generally increase with exit quality in this parametric range, the Qu and Siu-Ho correlation displays a nearly "quality-independent" behavior with a slight downward trend of the heat transfer coefficients with quality, having an MAE of 118% for inline and 129% for staggered. Parametrically, working fluid, heat fluxes, mass fluxes along with Prandtl and Reynolds number are within range of the Qu and Siu-Ho correlation. However, their high inlet subcooling and staggered square pin-fin geometry are substantially different from the current pin-fin array experiments. The McNeil *et al.* correlation has a trend similar to the data but substantially overpredicts the empirical results with an MAE of 363% for inline and 351% for staggered pin fin arrays. The overprediction by McNeil *et al.* could be explained by not only the larger 1 mm × 1 mm pin-fins used in their experiments, but also the R113 refrigerant working fluid that was used. The correlation with the best overall prediction capability for these empirical results is by Krishnamurthy and Peles with an MAE of 109% for inline and 144% for the staggered configuration. Once more, similar geometric differences exist between the circular, staggered pin-fin array used for the Krishnamurthy and Peles correlation and the present pin-fin geometry. Additionally, their inlet subcooling was much higher than used in the current work. These discrepancies are substantially

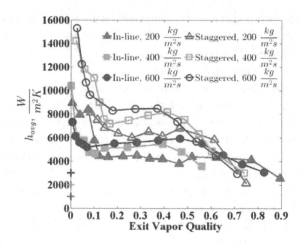

Fig. 12   Average heat transfer coefficient vs. exit quality for two-phase HFE-7200 in the inline and staggered arrays. "+" markings indicate single phase asymptotes.

beyond the $\pm 16\%$ measurement error and cannot be explained by experimental uncertainty alone.

HFE-7200, with a boiling point of 76°C at atmospheric pressure, entered the pin-fin array at 70°C, keeping the level of subcooling low to allow the exit qualities to be as high as possible. The same three mass fluxes of $200\,\mathrm{kg/m^2 s}$, $400\,\mathrm{kg/m^2 s}$, and $600\,\mathrm{kg/m^2 s}$, studied in the single-phase HFE-7200 experiments, were chosen for the two-phase experiments. Heat fluxes ranged from 1 to $36\,\mathrm{W/cm^2}$. Due to the low latent heat of the HFE-7200, the experiments spanned a broader range of exit qualities, exceeding 70% for all the experiments and reaching a maximum value of 90% for the inline pin-fin array operating at a $200\,\mathrm{kg/m^2 s}$ mass flux.

Inspection of Fig. 12 immediately reveals distinct differences between the two-phase HFE-7200 heat transfer coefficients and the water data in Fig. 11. Unlike the observed behavior with water, the HFE-7200 data reveal an approximately 50% improvement in the average heat transfer coefficient of the staggered array over the inline array, for much of the range of exit qualities. Most notable for both HFE-7200 array configurations is the initial sharp decline in the average heat transfer coefficient from the lowest exit qualities to about 10–15% followed by a plateauing or mild increase up to exit qualities of 40–50% where it reaches a local maximum. Finally, the average heat transfer coefficient deteriorates as the exit quality approaches 100%, possibly reflecting localized dryout in the pin-fin array. It should be noted that the two-phase heat transfer coefficients exceed that of the single-phase asymptote ("+" markings on plot) over the entire exit quality range, for all mass fluxes. The reader is informed that the ±16% measurement error bars were left out of Fig. 12 for clarity.

As expected, there is significant disagreement between the two-phase correlations in the literature and the HFE-7200 data. Especially of note are the multiple inflection points of the average heat transfer coefficient with exit quality which are not readily captured by 2 of the 3 available correlations. The Qu and Siu-Ho correlation in general overpredicts these empirical heat transfer coefficients with an MAE of 110.4% for the inline array and 59.32% for the staggered array. The Krishnamurthy and Peles correlation fails to capture the trend of the heat transfer coefficient with exit quality for HFE-7200, but has an overall MAE of 87.5% for the inline array and 93.6% for the staggered array. Since none of the available correlations were developed for HFE-7200, it is not surprising that they would not predict the current data well. In addition to the geometrical deviation of the current pin-fin arrays from each of the correlations, as mentioned in the previous section for water, the heat fluxes for HFE-7200 are particularly low for both arrays and out of the range of the correlations. Additionally, the exit qualities in the current HFE-7200 data substantially exceeded the maximum observed values for the studies leading to the correlations evaluated in the present effort. The maximum observed exit quality of 26% was in the experiments by Qu and Siu-Ho

The observed variation of the heat transfer coefficient with quality is reminiscent of the trends described previously in microgap flow boiling experiments by Rahim *et al.* (2011) Though it was suggested by Krishnamurthy and Peles that there may be flow regimes unique to micropin-fin arrays, such as bridge-flow (Krishnamurthy and Peles, 2008), the observed trend in this study is analogous to that occurring in microgaps and microchannels and may thus be explained by the general physics of two-phase phenomena in microchannels. Following Rahim *et al.* (2011), it can be expected that two-phase heat transfer coefficients will increase steeply from their single-phase values upon the initiation of nucleate boiling, for incrementally positive flow qualities, then decrease by transition to inter-mittent flow, as vapor "slugs" pass through the pin-fin array and induce portions of alternating thin film evaporation and local dryout at the wall and surrounding pin-fins. As the end of the slug-vapor intermittent regime and the onset of annular flow is approached, the heat transfer coefficient can be expected to plateau and then begin to increase as thin film evaporation becomes the dominant heat transfer mechanism and rising heat transfer coefficients result from thinning of the evaporating liquid film surrounding the pin-fins. Farther into the annular regime, a decrease in the heat transfer coefficient occurs, resulting from widespread local dryout of the liquid film. While the exact flow regime progression for pin-fin microchannels is as yet unknown, the similarity of the observed variation in the heat transfer coefficient with exit quality to that seen in microgap channels provides an initial basis for interpreting these empirical results.

As described in the current section, correlations available in the literature are unable to predict the current two-phase heat transfer coefficient data, especially over the broad range of exit qualities that were investigated. Therefore, it is important to develop a robust new correlation that can predict the performance of the inline and staggered arrays for both water and HFE-7200 with low average error. Since the Krishnamurthy and Peles correlation had the best overall performance, we will start with the form of their correlation and make a few key changes to improve it. First, the Nusselt number correlation by Short *et al.* used by Krishnamurthy and Peles was originally developed for large, air-cooled pin-fin heat sinks at laminar Reynolds numbers less than 10^3. Since good prediction accuracy for the current single-phase data was obtained with the Tullius *et al.* Nusselt number correlation using optimized shape factors in Sec. 3, these will be used in place of the Short *et al.* relation.

Next, the constant $\zeta = 1$ correction factor for the average heat transfer coefficient, will instead be replaced by an enhancement equation with exit quality and mass flux dependence. The equation will have five constants, $C_1 - C_5$. This will facilitate generation of the final correlation for average heat transfer coefficient by allowing adjustment of the shape of the curve for both pin-fin arrays over the entire range of exit quality. The form of the equation will be

$$\zeta = C_1 e^{C_2 x_e} + C_3 x_e^3 + \left(\frac{C_4}{G + C_5} \right)^{1/2} .$$

The form of this equation has a quality-dependent exponential function in the first term, an exit quality-dependent cubic function in the second term, and a mass flux-dependent function in the third and final term.

After using this new two-phase equation and selecting the constants $C_1 - C_5$ that minimize MAE for both arrays, the resulting prediction curves for deionized water are shown in Fig. 13. Since the experimental water heat transfer coefficients for the inline and staggered arrays were nearly the same,

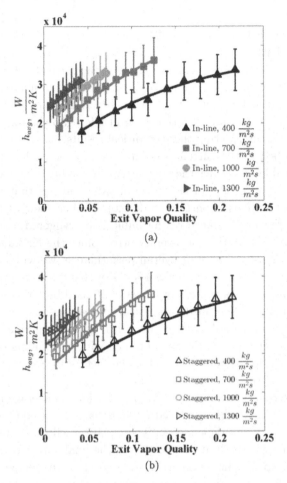

Fig. 13   New two-phase heat transfer coefficient correlation prediction for water in the inline array (a) and staggered array (b).

one set of constants were used to generate the equation. A remarkably small MAE of 2.44% was obtained overall for water.

For HFE-7200, two sets of constants — one for the inline and one for the staggered arrays — were optimized separately. The prediction curves are shown in Fig. 14. An MAE of 13.16% was obtained for the inline array and an MAE of 10.18% was obtained for the staggered array. A summary of the new correlation along with the constants used is given in Table 1.

Time variation of heat flux: Experiments under condition of time-varying heat flux were carried out, once again using the test section from

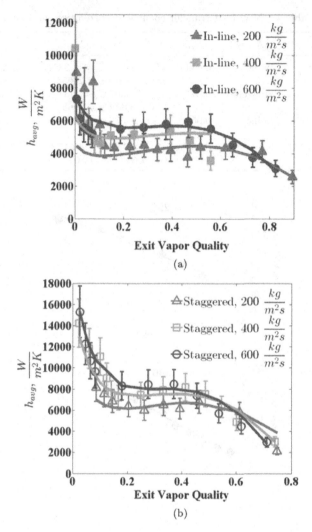

Fig. 14 New two-phase heat transfer coefficient correlation prediction for HFE-7200 in the inline array (a) and staggered array (b).

David *et al.* (2014) at two values of mass flux $m = 230 \, \text{kg/m}^2\text{s}$ and $m = 380 \, \text{kg/m}^2\text{s}$. The lowest heat flux was applied to maintain the steady state heated wall temperature in the range $T_W = \pm 2°\text{C}$. Then, during a time interval of 9–10 s, we increased the heat flux linearly with respect to time until the heated wall temperature was within a few degrees of $T_W = 50°\text{C}$. Data sets were recorded and averaged. Figures 15(a) and (b)

Table 1   New heat transfer coefficient correlation summary.

Fluid	Array	C_{Nu}	$C_1 - C_5$	MAE
Water	Inline	0.0495	$C_1 = -0.07$ $C_2 = 4.3$ $C_3 = 0$ $C_4 = 80$ $C_5 = 2965$	2.44%
Water	Staggered	0.0413	$C_1 = -0.07$ $C_2 = 4.3$ $C_3 = 0$ $C_4 = 80$ $C_5 = 2965$	2.44%
HFE-7200	Inline	0.054	$C_1 = 2.47$ $C_2 = -9.2$ $C_3 = -1.71$ $C_4 = 45$ $C_5 = 181$	13.16%
HFE-7200	Staggered	0.065	$C_1 = 6.0$ $C_2 = -14.15$ $C_3 = -3.63$ $C_4 = 45$ $C_5 = 88$	10.18%

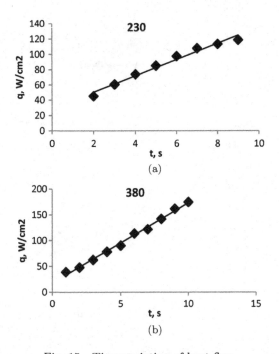

Fig. 15   Time variation of heat flux.

show time variation of heat flux at $G = 230\,\text{kg/m}^2\text{s}$ and $G = 380\,\text{kg/m}^2\text{s}$, respectively. The deviation of the measurements from the straight line cannot be distinguished within the uncertainty range.

Time variation of average heated wall temperature: One of the most important parameters of the heat sink is the temperature of the heated wall, T_W, often called the base temperature. The base temperature in electronic packaging is the reference temperature for all the electronic components

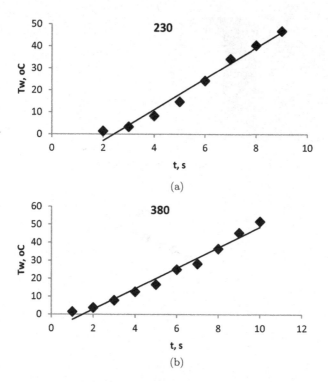

(a)

(b)

Fig. 16   Time variation of temperature on the heated surface.

attached to the base. Using this reference temperature, one can estimate the maximum junction temperatures and decide if a given component may be employed. Figures 16(a) and (b) illustrate the time variation of the average heated wall temperature at $m = 230\,\text{kg/m}^2\text{s}$ and $m = 380\,\text{kg/m}^2\text{s}$, respectively. The heat flux varied according Figs. 15(a) and (b). The wall temperature increases with time up to 46.8°C at $G = 230\,\text{kg/m}^2\text{s}$ and up to 51.6°C at $G = 380\,\text{kg/m}^2\text{s}$. It may be concluded that for known heat sink, mass and capacity and known time variation of heat flux, the value of mass flux may be chosen to keep the maximum reference temperature within a given range.

Boiling parameters under temporal variations of heat flux: Figure 17 shows the temperature field and histogram at fixed time instant of 10 s, and $q_{\text{max}} = 170\,\text{W/cm}^2$, $G = 380\,\text{kg/m}^2\text{s}$. The time- and surface-averaged wall temperature is 51.6°C, the maximum deviation from the average value does not exceed ±2°C. This value does not differ significantly from the value of ±1.5°C obtained under steady state condition.

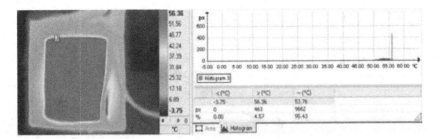

Fig. 17   Temperature field on the heater.

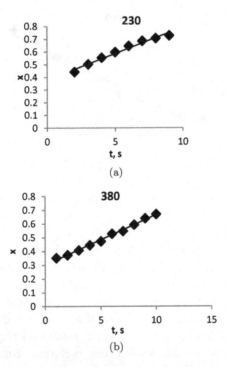

Fig. 18   Time variation of vapor quality.

Figure 18 shows time variation of vapor quality at the outlet of text section. Figures 16(a) and (b) show the variation of the heat transfer coefficient with vapor quality at the outlet of the test section for different values of heat flux. This quality was varied experimentally by linearly increasing heat flux during time of 9–10 s, while maintaining a constant mass flux.

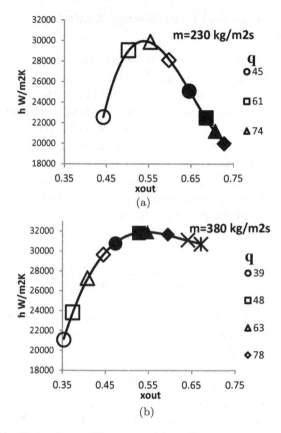

Fig. 19   Dependence of heat transfer coefficient on vapor quality.

Figures 19(a) and (b) show that up to vapor quality of about $x = 0.55$, the heat transfer coefficient increases with increasing vapor quality.

When the vapor quality exceeds the value of $x > 0.55$, the heat transfer coefficient decreases with increasing vapor quality. Such a behavior agrees with results reported above. Another noteworthy feature of the same data is the larger magnitude of the heat transfer coefficients compared to values obtained under steady state conditions. Under transient conditions, the dissipated power is absorbed not only by the working fluid but partially also by the material of the heat sink. For example, at $G = 380\,\text{kg/m}^2\text{s}$ and $q_{max} = 170\,\text{W/cm}^2$ the heat transfer coefficient was 25,000 and 31,000 W/m²K for steady-state and time varying conditions, respectively.

6. Least Material and Least Energy Analysis for CPV Cooling

The coefficient of performance (COP) is traditionally used to describe the cooling capability or heat output of a thermodynamic system in relation to the electrical or mechanical energy used to drive the cooling or heating process and serves as a basis of comparison for heat pump and refrigeration equipment. It is expressed as $\text{COP} = \frac{Q}{W}$ where Q is the heating energy or cooling output (kWh) and W is the energy input (kWh).

With a modest redefinition, this metric can also be applied to actively cooled CPV cells, taking the ratio of the useful electrical power generated by the cell (solar energy harvest or net solar energy) to the power consumed by the pump to cool the cell. The equation for COP used in this way will be given as

$$\text{COP} = \frac{E_{\text{PV}} - P_{\text{pump}}}{P_{\text{pump}}} = \frac{\text{Solar Energy Harvest}}{\text{Pumping Power}}. \tag{6}$$

This equation depends only on solar energy and pumping power and does not include energy from other sources. However, it could be modified to include other parasitic losses e.g. transmission line loss and power for the control electronics. In addition to accounting for the pumping power and parasitic losses, energy associated with the mining and refining of the raw materials, as well as the manufacture, transportation, and final assembly of all the various components and materials in a CPV system should be considered. Such an extensive energy analysis, cataloging, quantifying, and optimizing the energy content for each of these processes for all the components is beyond the scope of the present effort. Instead, this study will limit its attention to the embedded energy in the microcooler material and the required pumping power.

The total mass of the copper used in the fabrication of the microcooler determines the embedded energy content and has a direct impact on the performance of the cooling system. The material mass has associated formation energy for processing the copper, and additional energy is required for the further refinement or "fabrication" of that raw metal into its final form. Ashby (2009) found that 27 kWh/kg is the value of embodied energy for copper taking into account material, processing and recycling energy.

In the COP of Eq. (6), we will add the embedded energy to the pumping power in the denominator and convert the power terms to work terms by multiplying by the total lifetime hours of operation, t_L. The result is a total coefficient of performance (COP_T) metric defined as:

$$\text{COP}_T = \text{COP}\left(\frac{P_{\text{pump}}t_L}{P_{\text{pump}}t_L + 27{,}000\,\text{m}}\right)$$

$$= \frac{\text{Solar Energy Harvest} \cdot \text{Lifetime Hours}}{\text{Pumping Work} + \text{Embodied Energy}}. \qquad (7)$$

Although the COP_T metric was derived from the COP, it is distinct in that embedded energy is included to account for the energy required for the formation and fabrication of the copper as well as the lifetime energy of the pump. In addition, since multijunction cells are expected to last 25 years or more in a stable environment, and the solar industry is under pressure to increase cell lifetime to at least 30 years, in this analysis total lifetime t_L will be taken as 30 years assuming CPV operation for an average of 12 hours per day. It will be shown that COP and COP_T can be useful metrics to aid in identifying the system geometry that allows the most efficient use of mass and pumping power, while maintaining good cooling performance and high solar cell efficiency. Finally, it should be noted that the COP and COP_T are indirectly dependent on system parameters such as the fin geometry, flow rate, solar concentration, etc.

For the forthcoming analysis, a cell aperture area equivalent to the 28.8 mm × 9.6 mm base area of the pin-fin coolers will be assumed. This is a valid assumption since Spectrolab 40% efficient, triple junction CPV cells are available in multiple sizes, as small as 5.5 mm × 5.5 mm. Therefore, the 28.8 mm × 9.6 mm area could be considered as a cooling "module" of 3 or more CPV cells, which could then be used with other modules in a theoretical two-phase manifold cooling system. This concept is similar to the Solar Systems single-phase liquid cooling manifold design as described previously.

To round out the comparison, longitudinal-finned microchannels of similar geometry and aspect ratio to the inline pin-fin array, and a microgap cooler will be included in the model. A single channel microgap cooler is included in this comparison due to being the best of the longitudinal channels as found in Reeser *et al.* (2014). The microchannel cooler will have 31 channels with the same channel width and height of 153 and 305 μm, respectively. The microgap cooler will have 1-mm thick walls and a 1-mm thick base with a channel height of 305 μm. All coolers are assumed to have the same 1 mm thick base wall, along with a 50-μm layer of 63% Sn/37% Pb solder as the cell's thermal interface material. Working fluid for all simulations will be water.

Figure 20 is the solar energy harvest, which is the total power generated by the theoretical Spectrolab triple junction CPV module, minus pumping power, for a heat flux range from 20 to $165 \, \text{W/cm}^2$. Embodied energy is not included in the solar harvest analysis or Fig. 20. A constant flow rate of $33 \, \text{g/min}$ for the top plot, and $70 \, \text{g/min}$ for the bottom plot of Fig. 20 is assumed for each cooler in each of the respective plots. It is easy to see upon inspection of both plots that the pin-fin energy harvest is better for the pin-fin arrays than the microchannel and microgap coolers by 1–10 W, depending on the concentration ratio and flow rate. The difference between the inline and staggered arrays ranges from less than 1 W to 1 W with the inline array having a slight advantage in solar harvest.

For the low flow rate in the top plot of Fig. 20, the single-phase microchannel, the single-phase pin-fin coolers and single-phase microgap cooler are not able to provide cooling above 800 suns. Also, the two-phase microgap cooler cannot provide cooling above 1100 suns due to reaching CHF above this point. Further, both the two-phase pin-fin coolers, which are able to provide cooling to over 1600 suns, will generate 160 watts of usable power for our theoretical CPV module.

Shifting attention to the bottom of Fig. 20, we can see that the pin-fin arrays still facilitate the best solar power generation by the CPV module. However, due to the high flow rates in this case, the single-phase pin-fins are able to provide lower average base temperature and thus generate 10 more watts than the two-phase pin-fin coolers at an eqivalent concentration of 1500 suns.

The COP_T, which is defined in Eq. (7), is shown in Fig. 21 and includes the embodied energy of the copper microcooler. The highest COP_T of 8×10^4 is obtained by the single-phase microgap at 500 suns, which sharply increases up to this maximum value due to constant single-phase pumping power over increasing insolation. The two-phase cooling devices COP_T, however, are generally more constant. In the range shown, the two-phase pin-fin coolers stay near 10^4 over the entire range and are the most energy efficient microcooler for cooling above 1000 suns. In the bottom plot of Fig. 12, COP_T is substantially lower for all arrays due to the higher flow rate and thus higher pumping power. Once again, the inline single-phase pin-fins provide the best cooling, even up to 1700 suns, but does so only at a higher flow rate. Thus, at these higher heat fluxes or insolations above 1000 suns, the COP_T is higher — and therefore more energy efficient — when utilizing lower flow rate two-phase pin-fin cooling.

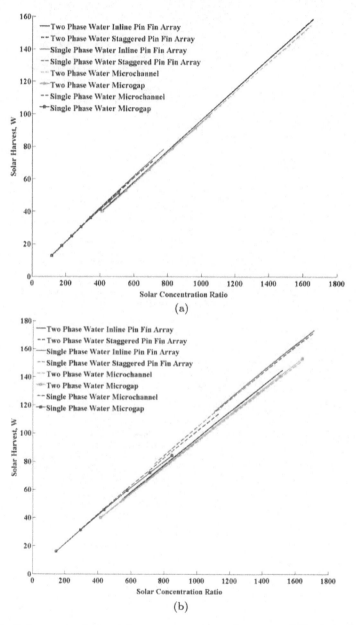

Fig. 20   Solar energy harvest for a constant mass flow rate of 33 g/min (a) and 70 g/min (b). Solar heat flux range from 20 to 165 W/cm².

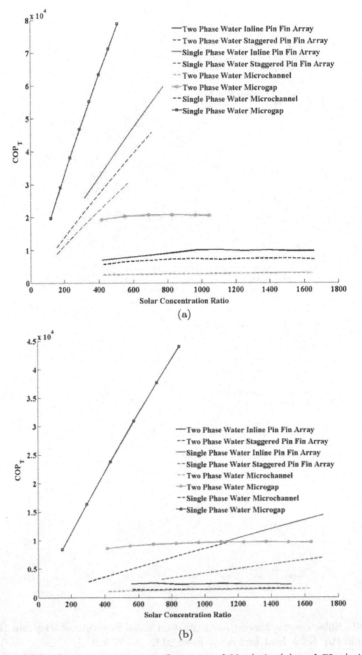

Fig. 21   COP$_T$ for a constant mass flow rate of 33 g/min (a) and 70 g/min (b). Solar heat flux range from 20 to 165 W/cm^2.

7. Conclusion

(1) Three unique micropin-fin arrays — staggered and inline — all of similar geometric proportion, were experimentally investigated in single- and two-phase flows for both deionized water and HFE-7200 working fluids up to exit qualities of 90%.

(2) single-phase and two-phase heat transfer coefficient behavior between water and HFE-7200 significantly differed over exit quality, with a distinctly increasing trend for water and a more "M" shaped curve for HFE-7200. We think it is reasonable to assume that the lower surface tension, lower liquid–vapor density ratio, and lower latent heat of HFE-7200 compared to water could play a significant role in the differing behavior of the two fluids in inline versus staggered pin fin arrays.

(3) Both single-phase and two-phase cooling can act as efficient CPV thermal management techniques; however, two-phase provides a distinct advantage over single-phase in terms of heat transfer coefficient.

(4) For high heat fluxes encountered at 1000 suns, and higher for high solar power generation at low flow rates, two-phase micropin-fins are the most energy efficient design for CPV cooling systems. For high flow rates and high heat flux cooling, single-phase pin fins provide the most energy efficient design choice. For low heat flux encountered at low concentration ratio, single-phase microgap maintains lower cell temperatures for the lowest parasitic pumping penalty. For both single-phase and two-phase cooling, inline pin-fin arrays are generally more energy efficient than staggered arrays.

(5) A technique for thermal visualization and determination of spatially resolved time series of wall temperature during flow boiling in a pin-fin microchannel heat sink was presented. The results of quantitative measurements, such as deviations of the surface temperature from time and space average values, are discussed. Results show that temperatures can be maintained with an uncertainty varying from $1.5°C$ at $q = 30\,\mathrm{W/cm}^2$ to $2.0°C$ at $q = 170\,\mathrm{W/cm}^2$. These results indicate that pin-fin microchannel heat sink enables to keep an electronic device near uniform temperature under conditions of steady state and time varying high heat fluxes.

(6) The heat transfer coefficient varied significantly with refrigerant quality and showed a peak at a vapor quality of 0.55 in all the experiments. At relatively low heat fluxes and vapor qualities, the heat transfer coefficient increased with vapor quality. At high heat fluxes and vapor qualities, the heat transfer coefficient decreased with vapor quality.

A noteworthy feature of the same data is the higher heat transfer coefficients attained under transient conditions, compared to values obtained under steady state conditions.

References

1. N. Yastrebova, "High-efficiency multi-junction solar cells: current status and future potential," (2007).
2. A. Luque and S. Hegedus, *Handbook Photovol. Sci. Eng.*, John Wiley & Sons Ltd, West Sussex (2011).
3. G. Landis, D. Merrit, R. P. Raffaelle and D. Scheiman, "High-temperature solar cell development", (2005).
4. M. Green, K. Emery, Y. Hishikawa, W. Warta and E. Dunlop, "Solar cell efficiency tables", *Prog. Photovol.*, Vol. 20, (No. 1), pp. 12–20, (2012).
5. P. Verlinden, A. Lewandowski, H. Kendall, S. Carter, K. Cheah, I. Varfolomeev, D. Watts, M. Volk, I. Thomas, P. Wakeman, A. Neumann, P. Gizinski, D. Modra, D. Turner and J. Lasich, "Update on two-year performance of 120 kWp concentrator PV systems using multi-unction III-V solar cells and parabolic dish reflective optics", in *Photovoltaic Specialists Conference*, (2008).
6. T. Ho, "Improving efficiency of high-concentrator photovoltaics by cooling with two-phase forced convection", *Inter. j. Energy Res.*, Vol. 34, (No. 14), pp. 1257–1271, (2010).
7. J. Tullius, T. Tullius and Y. Bayazitoglu, "Optimization of short micropinfins in minichannels", *Int. J. Heat Mass Trans.*, Vol. 55, pp. 3921–3932, (2012).
8. S. Krishnamurthy and Y. Peles, "Flow boiling of water in a circular staggered micro-pin fin heat sink", *Inter. J. Heat and Mass Trans.*, Vol. 51, pp. 1349–1364, (2008).
9. W. Qu and A. Siu-Ho, "Experimental study of saturated boiling heat transfer in an array of staggered micropin-fins", *Inter. J. Heat and Mass Trans.*, Vol. 52, (2009).
10. D. McNeil, A. Raeisi, P. Kew and P. Bobbili, "A comparison of flow boiling heat transfer in inline mini pin fin and plane channel flows", *Appl. Ther. Eng.*, Vol. 30, (No.16), pp. 2412–2425, (2010).
11. T. David, D. Mendler, A. Mosyak, A. Bar-Cohen and G. Hetsroni, "Thermal management of time-varying high heat flux electronic devices", *J. Electron. Packag.*, Vol. 136, (No. 2), (2014).
12. E. Rahim, R. R, J. Thome and A. Bar-Cohen, "Characterization and prediction of two-phase flow regimes in minature tubes", *Inter. J. Multiphase Flow*, Vol. 37, pp. 12–23, (2011).
13. M. Ashby, *Materials and the Environment: Eco-Informed Material Choice.* Butterworth-Heinemann, (2009).

14. A. Reeser, P. Wang, G. Hetsroni and A. Bar-Cohen, "Energy efficient two-phase microcooler design for a concentrated photovoltaic triple junction cell", *J. Solar Energy Eng.*, Vol. 136, (No. 3), pp. 03015–1:11, (2014).
15. A. Reeser, "Energy efficient two-phase cooling for concentrated photovoltaic arrays", Graduate Thesis, College Park, (2013).

Heat Transfer Characteristics in an Array of Micro-Pin-Fins: Single- and Two-Phase Flow

E. Pogrebnyak, B. Halachmi and A. Mosyak*

Faculty of Mechanical Engineering,
Technion, Haifa, Israel
*mealbmo@tx.technion.ac.il

The thermal characteristics of a laboratory pin-fin microchannel heat sink were empirically obtained for heat flux, q'', in the range of 1.5–48 W/cm^2 and mass flux, G, from 10 to 40 $kg/m^2 s$. Water and ethanol (C_2H_5OH, 96% pure) were chosen as the working fluids. Standard deviation of temperature distribution on the heat sink wall was less than $2.5°C$ in all experiments and less than $1°C$ in most. These results indicate that the pin-fin microchannel heat sink enables to keep an electronic device near uniform temperature under steady state conditions.

1. Introduction

Breakthrough in many cutting-edge technologies is increasingly dependent on the availability of highly efficient cooling techniques that are capable of dissipating a large amount of heat from small areas. For electronic cooling, it is desirable to maintain a low chip temperature in order to increase processing speeds and avoid chip burnout. One efficient method of heat removal is to use single- or two-phase cooling in mini- or microchannels. Microchannel cooling could be well suited for solving the problem, but the observed development of axial and lateral temperature non-uniformities on the surface of microchannel heat sinks is a significant problem for high heat flux dissipation. Nonuniform temperature distribution on the heated surface may be accompanied by a complex pattern characterized by spatial temperature variations that can produce potentially destructive thermal stresses along the interface between the chip and the substrate or heat sink. This is one of the key justifications for seeking nearly isothermal heat sinks.

Furthermore, a large temperature gradient is undesirable for the electronic performance since many electronic parameters are adversely affected by a substantial temperature rise. For instance, in electronic devices, electrical–thermal instability occurs within a high temperature region because the basic elements of electronic circuits have a switching time that decreases with increasing temperature (Qu and Sui-Ho, 2008; David *et al.*, 2014; Reeser *et al.*, 2014; Krishnamurthy and Peles, 2008; Peles *et al.*, 2005; Asadi *et al.*, 2014; Kandilkar *et al.*, 2012). For instance, Krishnamurthy and Peles (2008) experimentally studied flow boiling heat transfer in an array of staggered circular pin fins with diameter of 100 μm and 250 μm height. They found that heat transfer coefficient was moderately dependent on mass flux and independent of heat flux.

2. Fluid Loop

A fluid (water or ethanol), that was held in a reservoir, was moved by an adjustable pump. Next, the fluid went through a thermocouple/pressure transducer complex to measure its inlet temperature and pressure. Then, the fluid went through the heat sink apparatus, which included the micro-pin-fin heat sink. We used an electrical resistor and an electric power source to simulate the chip heat generation. We used an infrared radiometer in order to measure the resistor base temperature and an optical high-speed video camera to view the fluid flow from the other side of the heat sink. Upon leaving the micro-pin-fin heat sink, the fluid went through another thermocouple/pressure transducer complex to measure its outlet temperature and pressure. The fluid was then directed into a tank, later to be added to the reservoir attached to the pump. We weighed the fluid in the tank after a known period of time to calculate the flow rate.

3. Adiabatic Test Section

The test section, shown in Fig. 1, contains five elements: inlet manifold, inlet diffuser, base of housing, outlet diffuser, and outlet manifold. All parts, except the diffusers, were made of low thermal conductivity plastic material to minimize heat losses. The diffusers were designed to change the flow regime from circular flow to slot flow after the inlet manifold and from slot to circular flow before the outlet manifold. Due to the complex shape of the diffusers, they were manufactured of stainless steel in a complex wire EDM machining process. The diffusers have accurately reproduced dimensions.

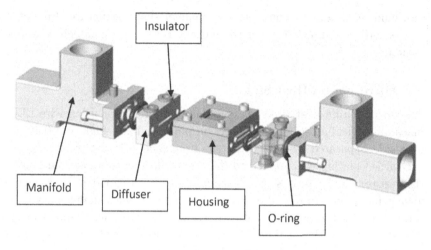

Fig. 1 Adiabatic test section.

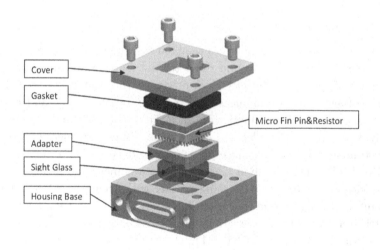

Fig. 2 Heated test section.

4. Heated Test Section

The pin-fin microchannel heat sink, with an attached 300 W resistor, sight glass, rubber gasket, and cover, comprised the test section, as shown in Fig. 2. The Sapphire sight glass was glued to the housing base. The pin-fin heat sink with its attached resistor is inserted inside the housing base. A special adapter is placed between the heat sink and the housing to eliminate any fluid leaks and to complete the channel shape structure. Finally, a rubber gasket is inserted under the housing cover, in a complete leak-free structure.

5. Micro pin-fin Heat Sink

The micro pin-fin heat sink was manufactured of copper using wire EDM machining, yielding square cross-section pin-fins, 1 mm high and 0.35 mm on a side, separated by a 0.35-mm gap from each adjacent pin-fin. Silver-over-nickel plating was applied to the heat sink base to allow soldering of the resistor to the heat sink. The microfins are designed in a staggered arrangement and are placed orthogonal to the flow, with the primary faces rotated 45° to the principal flow direction. A single microchannel block, placed in the middle of the test section, was used for the reported measurements and analysis. The resistor was heated by AC power supply and served to simulate the heat source.

6. Instrumentation

The temperature measurements of the fluid and the insulation in all blocks were carried out using 0.3 mm diameter T-type thermocouples with an accuracy of ±0.1°C. The electrical power to the heater was calculated using voltage and current measurements. The input voltage and current were controlled and measured with an accuracy of 0.5% and 1%, respectively. Pressures at the inlet and outlet of heat sink were measured by the 0–1000 mbar full scale absolute pressure transducers with an accuracy of 0.25% of full scale. The fluid mass flow rate was measured using a weight scale with an accuracy ±0.01 g and a timer with accuracy of 0.01 s. Cedip MWInSb Infrared System was utilized to study the temperature field on the electrical heater. This IR camera is suitable for temperature measurements in the wavelength range of 3.6–5.1 μm, with a sensitivity of 0.1 K and a typical resolution of 256 pixels per line. The measurement resolution was of 0.03 mm. Using the radiometer, one can obtain temperature in the point mode, quantitative thermal field in the line mode, and a temperature distribution for an area. We used universal USB data acquisition controllers to acquire the signals of thermocouples and pressure transducers with sample rate of 100 f/s. The controllers were connected to a computer and graphic user interface (GUI) was applied using commercially available software. For the finite element simulation, we used COMSOL multiphysics 4.3b software.

7. Experimental Procedure

Experiments were conducted with different flow rates, heat fluxes, and wall temperatures. While the flow rate was constant during each experiment, heat fluxes and wall temperatures varied. The wall temperature was measured using the IR camera, with the average temperature as an indicator. Measurements were taken with each mass flow/heat flux only after making sure the system has reached steady state conditions.

A series of flow rates was chosen; for each flow rate, we changed the heat flux to create a series of selected average wall temperatures making sure the temperature did not exceed a critical one. We repeated this procedure so each flow rate would have a series of approximately the same average wall temperatures. After using water as the working fluid, we repeated the experiments with ethanol. With ethanol, we had liquid phase experiments in the same wall temperature range as with water, whereas for the two-phase ethanol experiments the wall temperatures are higher.

8. Data Reduction

Data obtained from the voltage, current, weight, time, and temperature reading from both thermocouples and IR camera served to calculate the heat transfer coefficients and Nusselt numbers.

8.1. *Single phase*

The power transmitted to the fluid is

$$Q = \dot{m}C_P(T_{\text{out}} - T_{\text{in}}), \tag{1}$$

where $\dot{m}$ is the flow rate, C_P is the heat capacity of each fluid, and $T_{\text{in}}, T_{\text{out}}$ are the inlet and outlet measured temperatures, respectively.

The heat flux q_1 is calculated as

$$q_1 = \frac{Q}{A_1}, \tag{2}$$

where $A_1 = W \cdot L$ does not include pin-fin area, and L, W are the channel length and width, respectively. The heat transfer coefficient is calculated as follows:

$$h_1 = \frac{q_1}{(T_w - T_f)}, \tag{3}$$

where $T_f = \frac{T_{\text{in}} + T_{\text{out}}}{2}$, and T_w is the average wall temperature that is measured by the IR camera. Also, all the fluid properties are evaluated at the T_f temperature.

The Reynolds number for the flow is

$$\text{Re} = \frac{U d_h}{\nu}, \tag{4}$$

where d_h is the hydraulic diameter of the micropins (the same definition is also used by Kosar and Peles (2006) for microchannels with pin-fins), ν is the fluid kinematic viscosity, and U is the fluid average velocity.

The Nusselt number is

$$\text{Nu} = \frac{h d_h}{k_f}, \tag{5}$$

where k_f is the thermal conductivity of each fluid.

8.2.　Two phase

In the two-phase region, we can no longer use Eq. (1). The heat transferred to the fluid was calculated differently, namely

$$Q_{\text{tot}} = V \cdot I, \tag{6}$$

where V and I are the measured voltage and current supplied to the system, respectively.

The power which is transferred to the fluid is

$$Q = Q_{\text{tot}} - Q_{\text{lost}}, \tag{7}$$

where Q_{lost} was evaluated for each wall temperature. Energy balance was verified for each set of experimental conditions: the average temperature of the insulation was used to calculate the heat losses due to natural convection and radiation to the ambient. The heat loss during the experiments was found to be in the range of 3–5% of the applied heat.

We calculate the vapor quality x, which is

$$x = \frac{Q_{\text{boiling}}}{\dot{m} h_{fg}}. \tag{8}$$

8.3.　Experimental uncertainty

Errors depend on measurements of the following values: channel depth, H, channel width, W, channel length, L, pin height, b, pin width, w, inlet and outlet temperatures, T_{in} and T_{out}, mass flow rate, G, power transferred to the working fluid, Q, and heated wall temperature, T_W. The error in determining mass flux is due to errors of mass flow rate measurements. The error in magnitude of heat flux is due to the errors of Q, and heater dimensions, and heat losses. The error in the estimation of heat losses is due to the

Table 1 Experimental uncertainties.

Source of uncertainties	Symbol	Uncertainty
Channel width/length	W	0.3%
Pin width	a	1.5%
Pin height	H	1.5%
Mass flux	G	2.2%
Heat flux	q_1	4.2%
Heated wall temperature	T_w	0.4K
Vapor quality	x	9.8%
Heat transfer coefficient	h	11.6%
Inlet and outlet temperatures	T_{in}, T_{out}	0.2K

extrapolation statistic error. The uncertainty of measurements depends on bias limit, which is an estimate of the magnitude of the systematic error, and on precision limit, which is an estimate on the lack of repeatability caused by random errors. The uncertainty of components for error estimation was evaluated according to the Guide to the Expression of Uncertainty of Measurement.[a] The uncertainties in determining various parameters in this study are given in Table 1.

9. Results and Discussion

9.1. *Water*

In Fig. 3, the results for water as the working fluid are shown. It can be seen that for all mass fluxes the average wall temperature increases as the heat flux increases. Figure 4 shows behavior of the heat transfer coefficient as the heat flux increases with different mass fluxes.

In Fig. 4, one can see that the heat transfer coefficient does not have a decisive behavior as the heat flux increases. In all the water experiments, the difference between the maximum and minimum wall temperature was smaller than 6°C.

9.2. *Ethanol single-phase results*

The working fluid in this set of experiments was ethanol (C_2H_5OH). Figures 5 and 6 describe the same parameters as Figs. 3 and 4. The values in the legend on the right are the mass fluxes in units of $\frac{kg}{m^2 s}$. Similar to

---

[a] Guide to the expression of uncertainty of measurement 1995, International Organization for Standardization. Geneva.

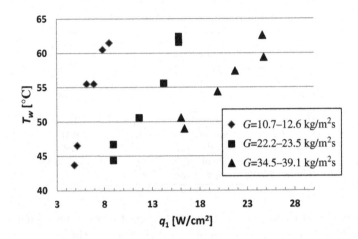

Fig. 3   Water. Dependence of wall temperature on heat flux.

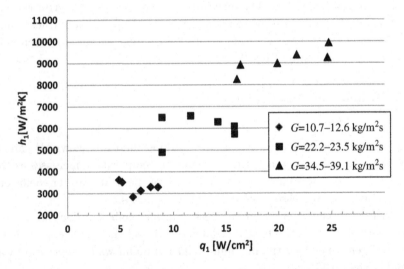

Fig. 4   Water. Head transfer coefficient vs. heat flux.

Fig. 3, the wall temperature increases as the heat flux increases for each mass flux.

In the three low-mass fluxes, the heat transfer coefficient increases as the mass flux increases; in the other four, higher, mass fluxes, it is more difficult to determine how the heat transfer coefficient behaves as the mass flux increases. In these experiments, the difference between the maximum and minimum wall temperature was lower than 5 K. For roughly the same

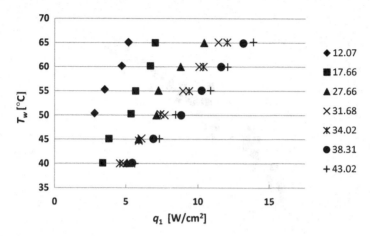

Fig. 5 Ethanol. Wall temperature vs. the heat flux for seven different mass fluxes.

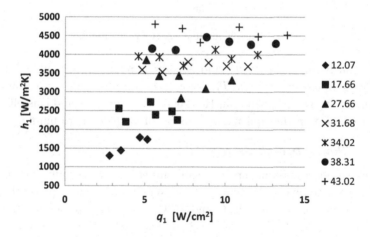

Fig. 6 Ethanol. Heat transfer coefficient vs. heat flux.

mass fluxes and same wall temperatures, water has a larger heat transfer coefficient than ethanol.

Figure 7 shows results obtained from water and ethanol in terms of the Nusselt number

$$Nu = 0.058 Re^{0.96} Pr^{0.4}. \tag{9}$$

The equation is close to that reported for water flow by Qu and Siu-Ho (2008):

$$Nu = 0.0285 Re^{0.932} Pr^{1/3}, \tag{10}$$

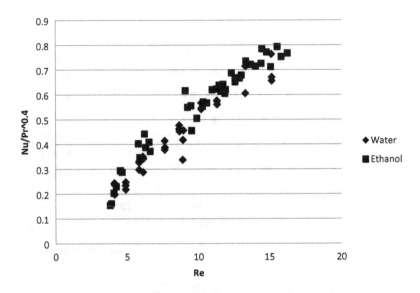

Fig. 7　Nusselt number vs. Reynolds number.

where the Reynolds, Prandtl, and Nusselt numbers are calculated at the average temperature denoted T_f. For correlations used in the present study and in the study by Qu and Siu-Ho (2008), the Reynolds and Nusselt numbers are based on the equivalent diameter of a single micro-pin-fin.

9.3. Ethanol two-phase results

In the two-phase experiments, the wall temperature was significantly higher than the single-phase wall temperature. While in the single phase the wall temperature was between 40°C and 70°C, in the two-phase part the wall average temperature was between 85°C and 95°C. In Figs. 8 and 9, the variations of the vapor quality, x, and the heat transfer coefficient increase as the heat flux increases for the six different mass fluxes (the mass fluxes are in kg/m²s).

The results indicate that in all mass fluxes the vapor quality increases as the heat flux increases. In the two-phase experiments, the wall temperature gradient increased up to a maximum of 10°C difference between maximum and minimum wall temperature measured. Unlike the single-phase results, in the two-phase results for all mass fluxes the heat transfer coefficient increases as the heat flux increases.

In the two-phase experiment, the ethanol does dissipate more heat than the single-phase water cases and has the same heat transfer coefficient for

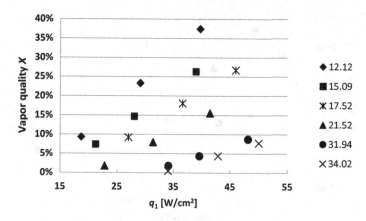

Fig. 8   Ethanol. Vapor quality vs. heat flux.

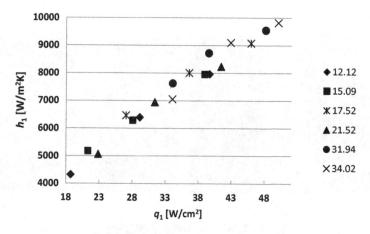

Fig. 9   Ethanol. Heat transfer coefficient vs. the heat flux.

roughly the same mass flux, but the wall temperature is significantly higher than in the single-phase experiment. We can expect that the water will dissipate more heat than the ethanol if the wall average temperature is the same. In Fig. 10, the variation of the heat transfer coefficient with the variation of vapor quality is presented.

Using a high-speed camera, we captured the behavior of the fluid inside the heat sink. Unlike the single-phase experiments, in the two-phase flow we managed to capture distinguishable changes in the flow over time. These changes can be seen in Figs. 11(a)–(c). The average flow direction in all

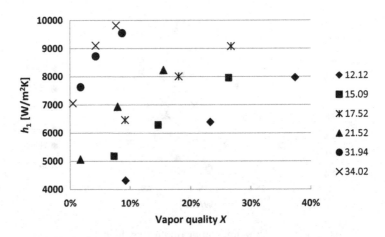

Fig. 10   Ethanol. Heat transfer coefficient vs. vapor quality.

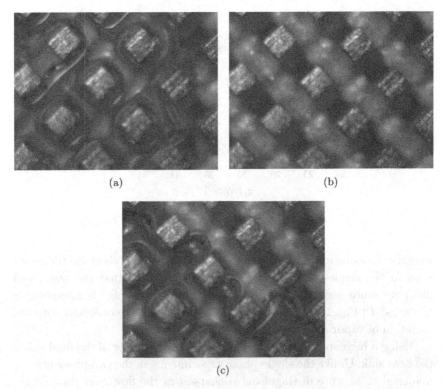

Fig. 11   Flow visualization. (a) $\tau = 12.5\,\mathrm{ms}$, (b) $\tau = 25\,\mathrm{ms}$, (c) $\tau = 37.5\,\mathrm{ms}$.

pictures is from the right lower corner to the left upper corner. The square objects are the pins. In Fig. 11(a), the gray area around the pins shows water flow; in Fig. 11(b), light strips between the pins show amount of vapor that moves in the flow direction; in Fig. 11(c), one can see that vapor moves also around the pins. Time interval between the frames is 12.5 ms. We can see from Figs. 11(a)–(c) that in a short period of time, a small area of the heat sink had both liquid and vapor flow. While this flow pattern is transient, the wall temperature was constant over time. It seems that this phenomenon is too rapid to have a noticeable effect on the wall temperature. This means the system can be used for two-phase cooling in these conditions.

10. Conclusion

The thermal characteristics of a laboratory pin-fin microchannel heat sink were empirically obtained for heat flux, q'', in the range of 1.5–48 W/cm^2 and mass flux, G, from 10 to 40 kg/m^{2}s. Water and ethanol (C$_2$H$_5$OH, 96% pure) were chosen as the working fluids. The heat transfer correlations for single-phase flow agree with results reported by Qu and Siu-Ho (2008). Standard deviation of temperature distribution on the heat sink wall was less than 2.5°C in all experiments and less than 1°C in most. These results indicate that the pin-fin microchannel heat sink enables to keep an electronic device in a state where, although the heated surface is not isothermal, the temperature variations are minimized.

References

1. W. Qu and A. Siu-Ho, "Liquid single phase flow in an array of micro-pin-fins-part 1: Heat transfer characteristics", *J. Heat Transfer*, Vol. 130, (No. 12), pp. 1–11, (2008).
2. T. David, D. Mendler, A. Mosyak, A. Bar Cohen and G. Hetsroni, "Thermal management of time varying high heat flux electronic devices", *Trans. ASME, J. Electron. Packag.*, Vol. 136, pp. 1–11, (2014).
3. A. Reeser, A. Bar-Cohen and G. Hetsroni, "High quality flow boiling heat transfer and pressure drop in microgap pin fin array", *Int. J. Heat Mass Transfer*, Vol. 78, pp. 974–985, (2014).
4. S. Krisnamurthy and Y. Peles, "Flow boiling of water in a circular staggered micro-pin-fin heat sink", *Int. J. Heat Mass Transfer*, Vol. 51 (Nos. 5–6), pp. 1349–1364, (2008).
5. Y. Peles, A. Kosar, C. Mishra, C. J. Kuo and B. Schneider, "Forced convective heat transfer across a pin fin micro heat sink", *Int. J. Heat Mass Transfer*, Vol. 48 pp. 3615–3627, (2005).

6. M. Asadi, G. Xie and B. Sunden, "A review of heat transfer and pressure drop characteristics of single and two phase microchannels", *Int. J. Heat Mass Transfer*, Vol. 79, pp. 34–53, (2014).
7. S. G. Kandilkar, S. Colin, Y. Peles, S. Garimella, R. F. Pease, J. J. Brandner and D. B. Tuckerman, "Heat transfer in micro channels-2012 status and research needs", *J. Heat Transfer*, Vol. 135, pp. 1–18, (2012).
8. W. A. Khan, "Modeling of fluid flow and heat transfer for optimization of pin-fin heat sinks", Ph.D. dissertation, Department of Mechanical Engineering, University of Waterloo, Canada (2004).
9. A. Kosar and Y. Peles, "Convective flow of refrigerant (R-123) across a bank of micro pin fins", *Int. J. Heat Mass Transfer*, Vol. 49, pp. 3142–3150, (2006).

Experimental and Numerical Investigation of Heat Removal by Microjets

Tomer Rozenfeld*, Yingying Wang‡, Ashwin Kumar Vutha†,
Jeong-Heon Shin‡, Gennady Ziskind*,§ and Yoav Peles‡,¶

*Department of Mechanical Engineering,
Ben-Gurion University of the Negev, Beer-Sheva, Israel
†Department of Mechanical,
Aerospace, and Nuclear Engineering,
Rensselaer Polytechnic Institute,
Troy, NY 12180, United States
‡Department of Mechanical and Aerospace Engineering,
University of Central Florida,
Orlando, FL 32816, United States
¶yoav.peles@ucf.edu
§gziskind@bgu.ac.il

In an ongoing collaborative work, experimental and numerical studies are performed to investigate the cooling ability of single-phase microjets of various configurations. Experimental studies include manufacturing and testing of microdevices. Three-dimensional numerical models of jet cooling are developed and implemented using commercial software ANSYS® Fluent. The simulations are used to reproduce the experimental conditions and to obtain detailed flow fields of the jet, heat flux, and temperature distribution on the impingement area, temperature distribution in the fluid, and heat transfer coefficients at the heated surface.

1. Introduction

Heat dissipation rates in processors and power electronics continue to increase, while electronic devices are constantly decreasing in size, making the pursuit of effective cooling solutions for these systems a challenging task. Impinging jets are a well-known and very effective cooling method that has been studied extensively at conventional scales (Zuckerman and Lior, 2006), and more recently at the microscale, showing very high heat transfer coefficients (Browne *et al.*, 2010). Usually, systems at the microscale include a channel that conveys the spent fluid while also serving as a confinement to the jet. This may have different, including positive, effects on the flow field and heat transfer.

An early study on the subject was performed by Zhuang *et al.* (1997), who experimentally investigated local heat transfer with liquid impingement flow in 2D microchannels. A significant enhancement of convective heat transfer in Reynolds numbers between 70 and 4807 was found. Sung and Mudawar (2008) investigated both experimentally and numerically a heat sink that combined a microchannel with an array of 14 circular jets. The results indicated that by reducing the channel height, the impinging effect was stronger on the surface, resulting in a lower maximum surface temperature, but accompanying by more noticeable temperature gradients. A recent work by Kim *et al.* (2015) numerically studied the effects of slot–jet length on the cooling performance of hybrid microchannel/slot–jet module.

In the current work, experimental and numerical efforts are undertaken to study the cooling potential of single-phase confined microjets of various configurations. Experimental studies include manufacturing and testing of microdevices with slot and round jets, where various working fluids are explored. In the numerical part, 3D models are developed to solve conjugate heat transfer for impinging jets in a microchannel. Simulations are used to complement experiments by obtaining detailed flow patterns of the jet, temperature, and heat flux distributions on the heated impingement surface and heat transfer coefficient estimations.

2. Physical and Numerical Models

The experimental devices are made of a 1-mm thick Pyrex and 400-μm thick silicon wafers. On the Pyrex wafer, several 100 nm thick resistance temperature detector (RTD) and a 1.5-mm $\times$ 0.4-mm heater are fabricated from titanium. Jet orifices are etched by deep reactive ion etching (DRIE) on a silicon wafer, which is attached to the Pyrex wafer through a 210-μm thick vinyl sticker. A 1.9-mm $\times$ 14.8-mm $\times$ 210-μm microchannel is formed by laser drilling into the sticker. The fluid enters through the nozzle inlet(s) and impinges on the heated microchannel lower wall, which is coated by a 2-μm thick silicon-oxide insulation layer, and exits through the two outlets located at both ends of the microchannel.

Figure 1 shows the schematics of the experimental microdevice, where the direction of the jet flow is denoted by blue downward-directed arrows.

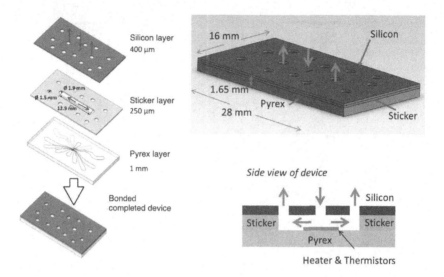

Fig. 1   Schematics of the experimental microdevice.

Fig. 2   The device and its inner part.

Figure 2 illustrates the device and compares its overall size to a common coin (US quarter).

Experiments are performed for varying flow rates of propylene glycol/water solution (40%/60%) flowing through a slot micronozzle, and gaseous nitrogen or HFE7000 flowing through several identical round micronozzles, with different heat fluxes, while temperatures of the RTDs are collected in order to study the impinging jet cooling characteristics. Figure 3 schematically depicts the heater and two different jet configurations. In both

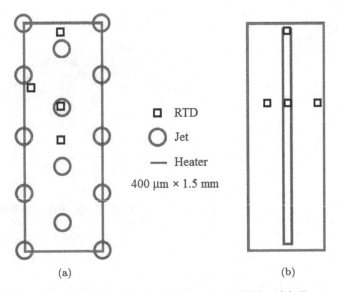

(a)   (b)

Fig. 3   Schematics of the heater, jet locations and RTDs. (a) Fourteen round jets, (b) single slot jet.

cases, the RTDs are denoted by black squares, which preserve their proportions as compared with the heater itself.

Three-dimensional numerical models of jet cooling are developed and implemented in this work using commercial software ANSYS® Fluent. The models are developed according to the experimental microdevice dimensions, materials, and conditions. The microchannel, nozzles, heater, and Pyrex substrate have the same dimensions as in the microdevice except for the microchannel length and outlets, in order to reduce the number of computational cells in unessential areas. In particular, the models do not include the microdevice circular outlets; thus, the modeled microchannel length is shorter and the outlets are located at the channel ends. Accordingly, the microchannel dimensions are 1.9 mm in width, 12 mm in length, and 210 μm height. The titanium heater on top of the 1-mm thick Pyrex substrate is modeled using thermal boundary condition of heat generation with dimensions exactly duplicating the experimental device. A 2-μm thick silicon-oxide insulating layer coated the entire upper surface of the Pyrex substrate, including the heater and the temperature sensors, and served as the channel lower wall interfacing the impinging jet.

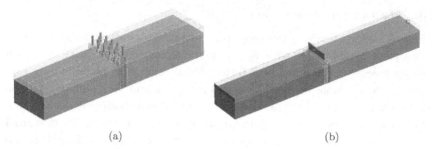

(a)   (b)

Fig. 4 Schematics of the model. (a) With several round jets, (b) with a single slot jet.

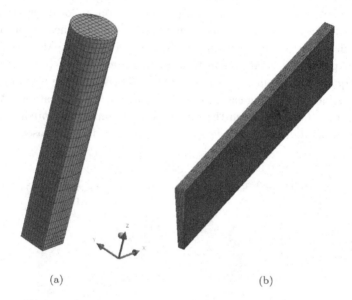

(a)   (b)

Fig. 5 Numerical model of the nozzles. (a) Round nozzle, (b) slot nozzle.

Figure 4 presents the schematics of the entire model, including the locations of the inlets (colored in green) and one of the outlets (colored in red). Figure 5 shows magnifications of the modeled nozzles: due to the manufacturing process, each round nozzle is circular at its top, 86 μm in diameter, and 54 μm $\times$ 54 μm square at its bottom. The slot nozzle has dimensions of 1.485 mm $\times$ 57 μm in its top cross-section and 1.47 mm $\times$ 34 μm in its bottom cross-section. The length of all nozzles is 400 μm.

3. Results and Discussion

Figure 6 shows temperature contours on the heater surface for a slot jet of propylene glycol/water with $Re_j = 232$ and power input of 0.9 W, which is equivalent to a nominal heat flux of $150\,\mathrm{W/cm^2}$. The black squares represent the locations of the RTDs, and the black and blue numbers are the experimentally measured and numerically predicted temperatures at those locations, respectively. A good agreement is found between the measured and numerically predicted temperatures with deviations of less than two degrees in most cases.

Figure 7 shows the velocity magnitude distribution for four out of fourteen simultaneous jets of nitrogen gas (N_2) with $Re_j = 633$ and Ma = 0.54. One can see that the flow field of a jet depends on its location within the jet array. Figure 8 shows temperature contours on the heater surface for 14 simultaneous jets of nitrogen gas (N_2) with $Re_j = 633$, Ma = 0.54, and power input which is equivalent to a nominal heat flux of $10\,\mathrm{W/cm^2}$. The black squares represent the locations of the RTDs, and the black numbers are the experimentally measured temperatures at those locations. A good agreement is found between the measured and numerically predicted temperatures, with deviations of less than two degrees in most cases.

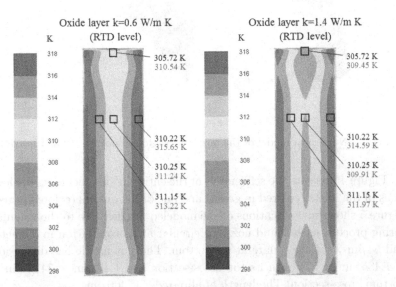

Fig. 6   Typical contours of temperature on the heater surface for $Re_j = 232$ and heat flux of $150\,\mathrm{W/cm^2}$.

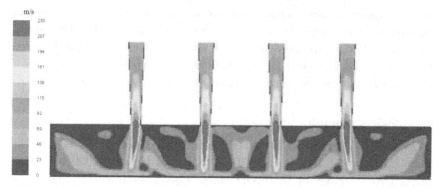

Fig. 7   Velocity magnitude distribution for four out of 14 simultaneous jets of nitrogen gas.

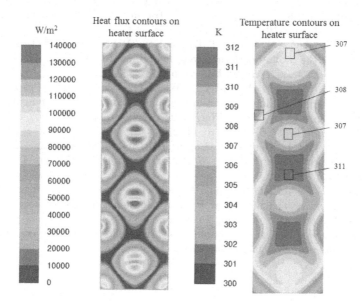

Fig. 8   Temperature contours on the heater surface for 14 simultaneous jets of nitrogen gas.

4.   Conclusion

The current work presented a study of single-phase slot and round jets impinging on a microchannel wall. Both experimental and numerical efforts are reported. The experimental effort is done in micromanufactured devices with desirable characteristics. The numerical models, developed and implemented using a commercial software ANSYS® Fluent, carefully reproduces the experimental geometry and parameters. A good agreement between the numerical and experimental results is demonstrated, enabling further investigation of the flow and heat transfer in similar microsystems.

Acknowledgements

This work was supported by the Office of Naval Research (ONR) and the Israel Ministry of Defense (IMOD), ONR award No. N62909-15-1-2025, under DARPA BAA 14-42; IMOD contract no. 4440715854.

References

E. A. Browne, G. J. Michna, M. K. Jensen and Y. Peles, "Experimental investigation of single-phase microjet array heat transfer", *J. Heat Transfer*, Vol. 132, No. 4, 041013 (2010).

C. B. Kim, C. Leng, X. D. Wang, T. H. Wang and W. M. Yan, "Effects of slot-jet length on the cooling performance of hybrid microchannel/slot-jet module", *Int. J. Heat Mass Transfer*, Vol. 89, pp. 838–845, (2015).

J. H. Shin, T. Rozenfeld, A. Vutha, Y. Wang, G. Ziskind and Y. Peles, "Local heat transfer coefficients measurement under micro jet impinging using Nitrogen gas (N₂)", *Proc. ASME 2016 Summer Heat Transfer Conference*, Washington, DC, USA, July 10–14, 2016.

M. K. Sung and I. Mudawar, "Single-phase hybrid micro-channel/micro-jet impingement cooling", *Int. J. Heat Mass Transfer*, Vol. 51, pp. 4342–4352, (2008).

Y. Zhuang, C. F. Ma and M. Qin, "Experimental study on local heat transfer with liquid impingement flow in two-dimensional micro-channels", *Int. J. Heat Mass Transfer*, Vol. 40, No. 17, pp. 4055–4059, (1997).

N. Zuckerman and N. Lior, "Jet impingement heat transfer: Physics, correlations, and numerical modeling", *Adv. Heat Transfer*, Vol. 39 (2006).

Printed in the United States
By Bookmasters